S88 Implementation Guide

Strategic Automation for the Process Industries

Darrin W. Fleming

Velumani A. Pillai

McGraw-Hill

New York   San Francisco   Washington, D.C.   Auckland   Bogotá
Caracas   Lisbon   London   Madrid   Mexico City   Milan
Montreal   New Delhi   San Juan   Singapore
Sydney   Tokyo   Toronto

Library of Congress Cataloging-in-Publication Data

Fleming, Darrin W.
    S88 implementation guide : strategic automation for the process
industries / Darrin W. Fleming, Velumani A. Pillai.
      p. cm.
    Includes index.
    ISBN 0-07-021697-5 (alk. paper)
    1. Process control.   2. Automation.   3. Production planning.
4. Manufacturing cells.   I. Pillai, Velumani A.
    TS156.8.F643   1999
    658.5'14—dc21   98-39559
                                              CIP

McGraw-Hill

A Division of The McGraw-Hill Companies

ISBN 0-07-021697-5

*The sponsoring editor for this book was Robert Esposito and the
production supervisor was Pamela A. Pelton. It was set in Century
Schoolbook by Ampersand Graphics, Ltd.*

Printed and bound by R. R. Donnelley & Sons Company.

To Kirk Joseph Fleming, November 29, 1968–July 7, 1990
This one's for you, bro!
—D.W.F.

To my loving parents, Rangarani and Arumugam
—V. A. P.

Contents

SECTION II OPERATIONAL DEFINITION

SECTION III DETAILED DESIGN

Preface

Most of the process industries are facing increasing external pressures that automation can help turn into competitive advantage. With fiercer competition, open markets, faster product cycles, increasing customer demands, and more regulation, companies are taking a hard look at their capital investments. Most companies underutilize their investment in process automation software and process control systems, and that investment can be extremely large. Quite often, they don't need to buy more equipment; they simply need to better utilize what they have.

When they do need to purchase equipment and/or software, they require nonproprietary open systems. What they are finding is that they do not have the resources or knowledge to optimize their systems. They need tools and assistance. When installing new systems, companies want to spend less on the equipment and end up with a better utilized, more flexible, and easier to use system.

The process control industry has responded by developing S88 "Batch Control." S88 is a standard developed by the International Society for Measurement and Control (ISA). Although focused on batch automation, S88 can be applied to any process automation. *S88 Implementation Guide: Strategic Automation for the Process Industries* teaches a methodology for developing and applying an automation strategy based on the foundation of S88. S88 does not describe how to implement automation; it simply outlines a set of models and terminology to be used. Readers of *S88 Implementation Guide* will learn how to apply a team-oriented approach to implement automation and achieve their desired results. The book covers adherence to OSHA's Process Safety Management (PSM) standards, the FDA's current Good Manufacturing Practices (cGMPs) standards, ISO 9000 series standards, and EPA standards.

Strategic automation is a team-oriented approach to automation. It teaches the models, methodology, and implementation to achieve true results with automation. It focuses on where the true value is derived—the process.

This book is written from the perspective of a team implementing automation within a company. It is divided into sections based on the phase of a project. This was done so that people with different roles within a company can read different sections. Through the book, you will follow the team as they apply the principles outlined to an example process.

Darrin W. Fleming
Velumani A. Pillai

Acknowledgments

Special thanks to my wife Nikki and my family for putting up with the many long hours needed to finish this book. Thanks are also due to Jolene for her many hours of editing and proofreading, the people at both PID and Intellution for their support, the ISA for enabling the SP88 committee to create such a powerful standard, Lynn Craig for his review and comments and for his encouragement, Tom Fisher for his review and comments, and Pat Schroeder for her review and comments. Jeff Anderson and Philip Schreiber also deserve a special thanks for their individual contributions to the content of this book.

Darrin W. Fleming

Many thanks to Uma for her encouragement and support, and her help in organizing the research material. She also made sacrifices and helped to maintain a good balance between work, the project, and home. I thank my friends and colleagues for their help during the project, for taking a great deal of interest, and in acting as sounding board when I needed one. I wish to acknowledge the help of several colleagues, notably Mike vanEpps and Dennis Regan, in identifying research sources, the constructive suggestions from Steve Mills and Rick Mol, and critical reviews by Venkat Ravoori and Dennis Regan. Special thanks to Necdet Kuruoglu for his vigorous support, thorough review, and creative and useful suggestions.

Velumani A. Pillai

Background
and Requirements

1

The Evolution of Process Automation

In order to appreciate where automation is today, it is helpful to look at how we got here. As with most aspects of technology, the pace of change is ever increasing. The process design cycle has been evolving for many years and has grown to be an efficient methodology for the design of a physical process. The following dialogue is an example (from the perspective of process control) of a typical design cycle for a process.

Typical Design Cycle

Patty* (a process engineer), develops a new processing technique for making a product. She draws a process flow diagram (PFD) showing her design, which includes the major processing equipment. She takes this drawing to an engineering design group where Larry (a chemical engineer) is assigned to the project as the lead engineer.

Larry works with Patty to refine the PFD and develops a conceptual design of the process. At this point, a design team is formed. The design team includes Larry, Patty, Mark (a mechanical engineer), Irene (an instrument engineer), and others whose project disciplines are less involved with process control.

The design team uses a conceptual design and flow diagram to create piping and instrument diagrams (P&IDs). Larry and Mark design the major pieces of equipment and work on the process layouts based

*Each character's name corresponds to the first letter of their role in the project.

on that equipment. Mark is responsible for the P&IDs, so he designs them based on the PFDs and process layout.

Once the P&IDs are near completion, the process is reviewed by a safety committee. The safety committee recommends a few alarms (which require instruments that weren't shown in the P&IDs) and interlocks. A hazard and operability (HAZOP) study is performed on the process. The HAZOP team recommends a few more alarms (also new) and interlocks. Patty takes another job within the company, and Larry assumes the role of process design.

At this point, the detail design of the process begins. Irene takes the P&IDs and determines the number and types of instruments required. She talks to a manufacturing representative to find out the plant's requirements, and then specifies a system. She begins writing specifications for each instrument, including type, size, range, special requirements, etc. Irene is also responsible for the control system. After ordering the instruments and valves, she starts developing the control strategy. The P&IDs show the measurement devices, control valves, motors, and control loops. After configuring the control shown on the P&ID, she configures the alarms and interlocks. Automatic start-up and shut-down are part of the project requirements, so she uses the sequential capability of the control system. Irene documents her sequences using flowcharts. Throughout the detail design, she directs all of her questions on how the process should run to Larry. Irene is now assigned to another project that needs her immediately.

The equipment, instruments, and control system arrive and are installed. A start-up team is created from the plant to oversee the commissioning and start-up of the process. Steve is assigned as the start-up coordinator. The start-up team checks the equipment, instrumentation, and control system installation and operation. They also write standard operating procedures (SOPs) for the process. A member of the start-up team has some process control experience, and so is assigned to work out the bugs in the system. The process starts up after some unexpected delays. As soon as the process starts making product, the start-up team is disbanded and the control system is turned over to the maintenance department.

The design team met the time and cost requirements. Everyone on the design team did a great job with the tools that were available.

Typical Results

Although everything went well on the project, there were some problems. The sequences to start and stop the process did not work. The documentation for how they were supposed to work was not adequate. Scattered comments throughout the code are not sufficient for someone to pick up where another has left off. The Start-up team practically had to reprogram the control system. At first, the control system wouldn't even make it past the interlocks in order to run the process. The start-up team had to work around many of the required interlocks to get the process to run. There were pumps that had interlocks that wouldn't allow them to ever start.

The control loops were tuned during start-up for steady-state conditions. These ran great during normal operation. Whenever the process was being started up, or a major upset occurred, the operators had to put them in manual. Eventually, they just kept them in manual and ran the process as if it were a manual process, except they ran it from the control room.

Whenever an upset occurred, the system generated a number of alarms. There were also nuisance alarms that tripped periodically for no apparent reason. The operators got to the point where they ignored the alarms and simply acknowledged them without looking at what they were indicating.

Whenever an interlock tripped, the control system lost control of the process, and the original upset caused upstream and downstream processes to shut down.

Each operator had his or her own way of operating, and thus there was an upset introduced at each shift change. Each operator was extremely efficient at parts of the process, and inefficient at others. One shift even went so far as to disable every sequence and controller.

Changes to the system were extremely difficult. There was little documentation generated after the original design, and much of what was available was changed during start-up. The plant safety department asked the start-up team for documentation of operator training for the Process Safety Management file, but unfortunately none existed.

What Went Wrong?

Victor, the Vice President of Manufacturing, decided that it was time to find out what was going on. This was not the first seemingly successful project to produce less than desirable results. He knew that it was not the fault of the people. There had been different people on many different projects and they had produced similar results. Based on that, Victor assumed that there must be gaps in the tools and methodologies that were being used. He decided to form a team to investigate the tools and methodologies and propose improvements. The team consisted of the same people on the original project with some additions. Once they made the recommendations, they were to utilize the new tools and methodologies on the design of a new process. Since it is usually easier to objectively evaluate a situation from the outside, he hired Allen, an automation Consultant, to lead the team.

During their first meeting, tensions were high. Everyone initially assumed that this was a witch hunt, and that someone would take the fall. Victor was quick to assure everyone that this wasn't true. He said he knew that there was a systemic problem with the engineering approach used for the project, and he challenged the team to lead the company to a new level of technological advancement. He said their goal was to find and develop the tools and methodology to gain a competitive advantage through automation. He promised they would have the time and support to achieve their goal.

They agreed up front that they must maintain mutual respect. With this as their common bond they began their journey. Larry started the discussion.

LARRY: "Victor, you said that you want us to gain competitive advantage through automation, but we are working with a process control system. Are automation and process control the same thing?"

VICTOR: "I guess I'm not sure. Allen can you help us understand?"

ALLEN: "There are some significant differences. How about if I talk about them later. Let's start by talking about the last project."

Steve then recalls that the safety department really came down on him when he couldn't give them documentation for operator training.

STEVE: "They told me that documentation and training are two of the fundamental issues we need to address in order to meet regulations. In the future, we need to plan for them up front."

LARRY: "Does anyone really know why we automated the process anyway? What were we trying to gain? It sometimes seems like this automation thing is just techno-lust."

They were all surprised to find out that, although some had ideas, no-one really knew what the objective was on that particular process.

ALLEN: "Unfortunately Larry, the result of not having a clear understanding of the objective of automation is similar to the result of applying technology for the sake of technology."

PATTY: "Let's start an issue board (Figure 1.1) to keep track of our thoughts. Write all of these things on our issue board. We need to know why we are automating."

At that point, the ideas slowed down, so Allen decided it was time for him to facilitate the conversation.

ALLEN: "Who designed the automation?"

IRENE: "I guess I did most of it."

ALLEN: "In retrospect, do you think you knew everything you needed to know about how the process should operate?"

IRENE: "Not even close, I guess others should have been more in-

> • Documentation and training need to
>   be defined up front
> • Need clear understanding of
>   automation objective

Figure 1.1 Issue board 1.

volved in the design. For one, it would have been nice if Patty could have been more involved. She knew how she wanted the process to run."

STEVE: "Someone from the plant should have been involved also. Maybe an operator, or someone who had been through a few start-ups."

LARRY: "So we need to have the right people involved. There's another issue!"

STEVE: "It seems to me that with all of these fancy new computers, we should be able to do things better than we ever could manually. Unfortunately, every process, I've worked on operated like a manual process except we used the computer to interface with it. What are we gaining by automating what we were already doing?"

MARK: "What's that, four issues? We're really rolling now!" (See Figure 1.2.)

ALLEN: "Who has heard of object oriented technology or object oriented programming?" (Most nodded.)

LARRY: "Sure, but what does it have to do with this project?"

ALLEN: "I'll get back to your question, Larry. How many have heard of S88?" (Everyone looked puzzled.)

IRENE: "Isn't that one of the ISA standards like Fieldbus or something?"

ALLEN: "Close. S88 is an ISA standard on batch control. Although most of your processes are continuous, many of the concepts can be applied equally as well. Now back to your question Larry, object oriented technology is a way of looking at the world. S88 is based on an object view of the world. I don't want to bog us down on this now, but I think that it is important to learn in what direction the rest of industry is going."

Figure 1.2   Issue board 2.

LARRY: "If you say so, then I'll add it to the list."

Irene pointed out that P&IDs did not adequately show how a process should operate.

IRENE: "A P&ID is a static picture—a time slice—of the process. We need a way to show a dynamic process." Everyone agreed that a drawing of the process was probably not the answer.

LARRY: "Since I'm up, I guess I'll continue being the scribe" (See Figure 1.3.)

STEVE: "I have something else. We need a better way to define and deal with interlocks. There were pieces of equipment that wouldn't even run because of interlocks."

IRENE: "I agree, I wanted to document them somehow, but I didn't know how."

LARRY: "Got it."

MARK: "I have a question. I was in the plant the other day and I saw an operator acknowledging alarms without even looking at them. Why does that happen?"

STEVE: "I take it you've never been the only one in the control room? Most of the alarms are meaningless. They tell you things like there is 'low flow' when you just turned off a pump." (See Figure 1.4.)

ALLEN: "I have a suggestion. I have some information on how automation has evolved, the differences between automation and process control, and some philosophies that I live by when implementing automation. So far, all of your comments have fit into one of these categories, plus there are a few others that I'm sure will come up eventual-

- Documentation and training need to be defined up front
- Need clear understanding of automation objective
- The right people need to be involved in the design
- Automation needs to improve on manual process
- Follow industry standards and trends
- Need better way to define the operation of a dynamic process

Figure 1.3   Issue board 3.

- Documentation and training need to be defined up front
- Need clear understanding of automation objective
- The right people need to be involved in the design
- Automation needs to improve on manual process
- Follow industry standards and trends
- Need better way to define the operation of a dynamic process
- Need better way to define and deal with interlocks
- Need to manage alarms so that they are meaningful

Figure 1.4   Final issue board.

ly. How about if I share them with you as a primer for our next meeting?" Everyone wanted to see the information. [Allen's account of the history of automation follows.]

Brief History of Automation

Process control has evolved throughout the ages. As early as 300 BC in Greece, mechanical devices were designed to regulate liquid level. The mechanical devices performed tasks that would otherwise be done by a person. The development of pneumatic control allowed process conditions to directly affect control actions (i.e., a tank-level measurement could change the position of a control valve). Any measurement that could be converted into an air pressure signal could be used to control a process.

As is well known, measurement alone is not enough to determine what needs to be done to a process. An algorithm has to be developed that calculates the control action necessary to bring a measured variable back to its setpoint.

The first algorithms were on/off feedback loops similar to those in a home heating system. These were found to be unstable and, in many cases, incapable of maintaining a target (setpoint). Eventually, single variable control evolved into the proportional, integral, derivative (PID) loop. A PID control scheme takes into account the amount the measurement is away from the setpoint, the length of time it has been away from the setpoint, and the rate at which the measurement is changing in order to calculate an appropriate control action.

As processing has become more sophisticated, so have the requirements placed on control strategies. Ratio control, cascade control, and feedforward control are examples of more elaborate control schemes. All of these control schemes are based on controllers that were designed to use one process measurement to determine the control action to take. They were usually designed to do what the operator was capable of doing manually. Advanced control strategies are, in most cases, groupings of single variable controllers arranged in a way to achieve the desired result.

Single variable controllers have a difficult time dealing with process dynamics. These controllers are inherently designed for steady-state conditions. As the dynamics of the process change, the responses to control actions change. A controller is adjusted to control at a given set of process conditions, but as the dynamics change, the controller is less able to maintain the stability of the process. Even though a controller can be adjusted to regulate under various conditions, in many cases the proper information is not available in a timely enough fashion.

As controllers became more high-tech, they became more capable and sophisticated. Self-tuning controllers, for example, are capable of adjusting their tuning parameters based on process dynamics. While this was a major step forward in dynamic processes, such controllers are still merely reacting to the dynamics of the measurement as opposed to predicting the dynamics based on the state of the entire process.

More recently, computers have been applied to the task of controlling processes. A computer's capability to monitor a multitude of variables and take many actions simultaneously has brought newfound capability to the process control industry. Sophisticated programs have been written to perform sequences of actions based on triggers (events). These events can emanate from the process, an operator, or other sources of information. Added to this, high-speed communications has resulted in a stunning increase in the number of possible ways in which a process can be controlled.

Automation versus Process Control

ALLEN: "I think it was Larry who asked if there was a difference between automation and process control. I personally believe there are some significant differences, but I'd like to see what the rest of you think. Here are my thoughts on the subject."

There are some key differences between automation and process control (classical control). For instance, classical control is typically designed to control a process during steady-state operation, whereas automation is designed to manage the process at all times. Classical control is designed to run the process in one and only one way; automation is designed to run the process based on what is necessary to make product. Classical control is designed around single variable controllers, automation is designed around modules.

Sequential control is the sequencing of events based on certain triggers and events. Although automation contains both classical control and sequential control, it coordinates the activities of the entire process at a much higher level.

Classical control is an integral part of an automation strategy. The control algorithms and advanced control schemes that have developed over the years are extremely efficient at what they are designed to do. So an automation strategy cannot be implemented without classical and sequential control. (See Table 1.)

TABLE 1.1   Automation versus Classical Control

	Classical control	Automation
Definition tool	Drawings	Database
Point of view	Loop	Process
Program alignment	By product	By Process
Programming method	Procedural	Object oriented
Operator interface	Control based	Process based
Exception handling	Stop everything	Minimize upset
Alarms	Sporadic and unmanaged	Managed by automation
Design criteria	Steady-state	Dynamic conditions
Engineering approach	Tactical	Strategic
Time base	Past/Present	Future
Level of definition	Low	High
Operator's job	Operate	Manage and optimize

Process Automation

Automation is the act of managing the process in all circumstances. Automation is the coordination of activities within the entire process, not the sequencing of events within an isolated portion of the process. Automation responds to process dynamics, whether they are measurements, interlocks, or operator actions, etc. In effect: automation manages the process.

When an event occurs that could constitute an emergency, automated processes and classical control systems respond in very different ways. A system based on classical control responds to the upset in a predetermined fashion. An automated process takes the same signal, and by knowing the condition of the entire process, takes the appropriate action (trying to minimize the upset) and eliminates the unsafe state. If the signal is a false warning that is only appropriate in other circumstances, it will ignore it. An automated process control system provides all of the information for the next level of automation, an intelligent control system. This is not yet at the level of artificial intelligence (AI), although it does provide all the appropriate information for an AI system to make inferences.

Classical control views the process as individual bits and pieces. Automation allows the entire process to be viewed as a whole. This holistic view of the process is available only because a modular or *object-oriented* approach is taken to the process organization. Since the individual modules are self-contained, the details of their operation can be ignored and they can be viewed relative to the other modules. Automation, by definition, manages the entire process during all transitions and in all states. This allows the controllers to be adjusted by the automation based on the condition of the process.

True automation allows different classical control strategies to be used during different process conditions. This can be as simple as using different tuning parameters during start-up than during steady state or it can be as intricate as using an entirely different strategy, with different loops and structure.

Automation has a strategic focus on the future direction of the process and products, because it enables your process to help you realize your business *strategy*.

True automation will redefine an operator's job. Operators will no longer perform process operation tasks; they will manage and improve the process. In a process using only classical control, operators will typically be operating the process, not improving it. They will be placing the controllers in automatic or manual, changing the setpoints, opening valves, operating the process in bits and pieces, not as a whole.

Let us look at a process in terms of flexibility, capability, and complexity (Figure 1.5). A totally manual process has as much flexibility as the process will allow. The capability of the process is limited by what people can accomplish. In order to add capability, more people need to be added. A single operator can only do so many things at a time. A totally manual process is as simple as can be in terms of engi-

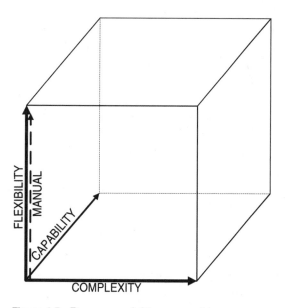

Figure 1.5 Process capability—manual.

neering and maintenance. We will use a manual process as the base-line for both capability and complexity.

Ideally (Figure 1.6), we would like to achieve increases in capability without losing flexibility. Unfortunately, that hasn't been the case. As we added technology to processes, we exponentially increased the complexity (Figure 1.7). In the past, adding computer systems to process control resulted in a loss of flexibility. The tools weren't available to achieve the same level of flexibility for computer controlled process as for manual ones. The programs were typically written for an individual product. Therefore, when a product needed to change, the program needed to change.

Through adoption of the practices and terminology of S88, you can bring process control closer to the ideal expectations of the end user, that is, highest flexibility. This requires discipline. Without it you may add capability, but usually at the expense of flexibility and simplicity. Good definition, a modular approach, and adherence to standards not only reduces complexity but increases the flexibility of the process (Figure 1.8). In other words, ***ultimate flexibility is achieved through enforced rigidity.***

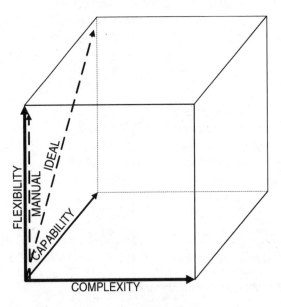

Figure 1.6   Process capability—ideal

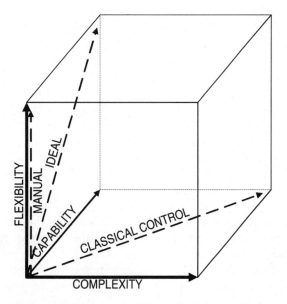

Figure 1.7   Process capability—classical control.

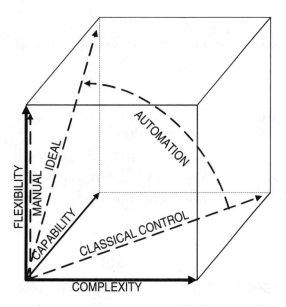

Figure 1.8   Process capability—automation.

Key Elements of Successful Automation

There are some concepts that should be remembered when implementing an automation project (Figure 1.9). We will discuss each of these in more detail as we progress.

IRENE: "I certainly understand the one about sticking with it until it works. I've never been able to carry a project through start-up. I always get pulled off onto another project."

STEVE: "Me too. Start-up teams are always disbanded before the bugs are really worked out of a process."

Summary

A major paradigm shift is required to achieve the step change results you desire. Stop thinking of the computer that controls your process as a process control system. It is a tool that helps you make product. It is no different than your mixer, your reactor, or your distillation column. Use the tool in creative new ways and you will gain a competitive advantage. Don't fall into the "it wasn't designed to do that" mindset. A wire wasn't designed to carry your voice but, luckily, someone forgot to inform Alexander Graham Bell of that.

The entire automation design cycle will be outlined in later chapters, including some of the philosophical changes and changes to the tool kits that must be considered.

- Know your Automation Objective
- Solicit your Experts
- Simplify and Optimize
- Focus on Adding Value
- Apply Modular Manufacturing Principles
- Make your Process Communicate
- Define! Define! Define!
- Retrain and Realign
- Stick with it until it Works
- Continuous Improvement

Figure 1.9   Key elements of successful automation.

2

Project Methodology

VICTOR: "After the last meeting, and based on your recommendations, I think a few more people need to get involved in this team. There are a lot of issues to discuss, and we need to have all the right people involved to get the job done. Let's go around the room and introduce ourselves and our roles. I'm Victor. I'm the Vice President of Manufacturing."

PATTY: "I'm Patty. I'm a Process Design Engineer."

LARRY: "I'm Larry. I'm a Project Manager out of Central Engineering. I usually perform the role of Lead Engineer on the design of new plants."

MARK: "I'm Mark. I'm a Mechanical Engineer in Central Engineering."

IRENE: "I'm Irene. I'm an Instrument Engineer out of Central Engineering."

STEVE: "I think I've worked with everyone in the room, but I'm Steve. I'm a Start-up Coordinator in the plant."

Owen, who was sitting in the back of the room with his arms crossed, simply said, "Owen. Operator."

DONNA: "My name is Donna and I work in Product Development. I interface between research and manufacturing on the development of new products."

ROBERT: "I'm Robert. I'm the Manager of Regulatory Compliance."

ALLEN: "I'm Allen. I'm a Process Automation Consultant."

VICTOR: "Allen and I spoke after the last meeting, and he has some information that he would like to share with us."

ALLEN: "First, I'd like to find out what everyone thinks a methodology for process automation should consist of or what problems it should solve."

LARRY: "It has to have a way to control project costs and manage changes."

PATTY: "Sure Larry, cost is important. But at some point in time we need to make sure that we are improving our processing technology."

IRENE: "We need to have a mechanism to get the right people involved early in the project. We need to end up with a better definition of how the process operates so that the person writing the control code knows what it should do."

DONNA: "It would be great if it could accommodate future product needs so that it wouldn't take three months to change the software to make a new product."

MARK: "We need to make sure that the operational strategy is well thought out, so that major process changes don't need to be made late in the game."

STEVE: "All of these things have been mentioned, and no-one has even said anything about the automation working to make product."

ALLEN: "Excellent point, Steve. We always need to remember that we are here to make product. Unless anyone else has any other items to add to the list, let me outline the 'Strategic Automation' methodology. This is a methodology that I have found to be very successful and I think it will address most of your concerns."

Automation Methodology

For a given process, the automation lifecycle may not match the process lifecycle. An existing process with no automation may be automated without any process changes. Typically, a process will outlive the automation. There may be several generations of automation implemented during the life of a process (Figure 2.1).

Strategic automation is a project methodology for implementing automation. It was designed to meet the needs of automation and meet current regulatory requirements (Figure 2.2). The strategic automation methodology has six phases (Figure 2.3) and several key attribut-

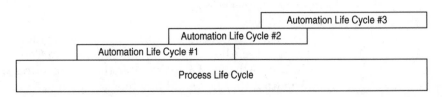

Figure 2.1   Life cycles.

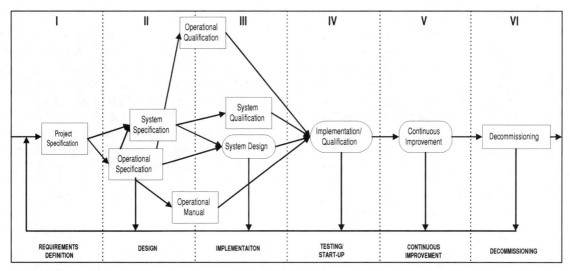

Figure 2.2   Strategic automation methodology.

es. The automation engineering effort is heavily front-end loaded. Change control is designed into the entire methodology. Note that there is nothing called "as-builts." This is because at any given time, the current version of documents will be able to describe the state of the process automation. Any change is evaluated against the impact on all previous phases of the project and is fully documented before it is implemented. Finally, testing of the installed system is part of the overall automation design. As the process is being designed, the critical parameters are being defined along with the protocol for testing and the acceptance criteria for each parameter.

I.   REQUIREMENTS DEFINITION
II. DESIGN
III. IMPLEMENTATION
IV. TESTING/START-UP
V. CONTINUOUS IMPROVEMENT
VI. DECOMMISSIONING

Figure 2.3   Methodology phases.

Phase I—Requirements Definition

During Phase I, the automation objectives are defined (Figure 2.4). The entire team must understand the objectives and how their roles contribute to the entire project. An automation project can provide many benefits, but if everyone doesn't understand the purpose of the project, different people may be working towards different objectives. When this occurs, few of the benefits are realized. In the end, the project must be measured against the defined objectives. Remember, "What gets measured gets managed." A clear and concise scope document needs to be written during this time. A scope document is a brief functional description of the process.

Phase I Objectives

- Write a scope document for the process
- Develop measurable, specific, and achievable objectives for the process
- Develop measurable, specific, and achievable objectives for the automation that is aligned with the process objective
- Define the metrics against which the project will be measured

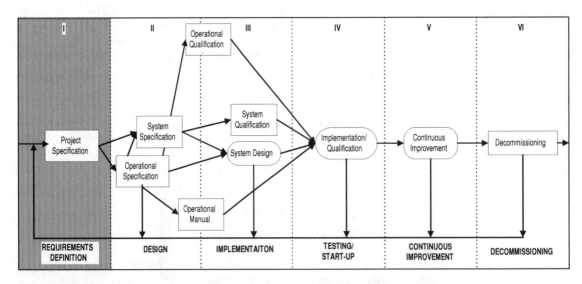

Figure 2.4 Methodology I.

Phase I Flow

- All key people involved
- Begin with the end in mind
- Phase I must be complete and approved prior to the start of Phase II

Phase I Deliverables

- Schedule
- Scope document containing approved objectives

Phase II—Automation Design

During Phase II, the operation of the process is defined (Figure 2.5). The operational specification is written in platform-independent terms, that is, it does not define how the control is to be programmed. The operation is also defined based on process capabilities. It is not defined based on product needs. The ISA S88.01 standard (which we will discuss later) makes a clear distinction between process and product, or in other words, between equipment and procedure. Unless automation is being applied to an existing process, the P&IDs should not be designed until Phase III. Phase II should be done with PFDs.

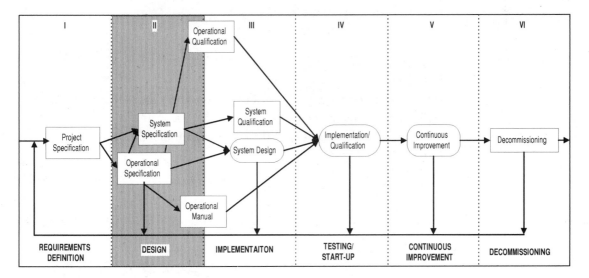

Figure 2.5 Methodology II.

Phase II defines both how the P&IDs should be organized and the measurement and output needs of the process. The systems and software that are to be used in the project need to be defined by the *end* of Phase II. Hazard analysis and operability studies (HAZOPs) are generally conducted during Phase II and Phase III, when all the inputs are available.

Phase II Objectives

- Simplify and optimize the process
- Modularize the process
- Define operational requirements that will be a part of operational specification
- Define alarming strategy
- Define process communication to and from the process
- System requirements that will be a part of system specifications
- System selection from system requirements

Phase II Flow

- Have all key people involved
- Define the physical model in a top-down manner
- Define the procedural control for each module
- Define the alarming strategy
- Approve all specifications

Phase II Deliverables

- Written description of the standard operating procedures (SOPs).
- Physical model of the process
- Approved operational specification for each module
- System specification
- Alarming strategy
- Roles and responsibility
- Qualification plan

Phase III—Automation Implementation

During this phase, the P&IDs are being developed unless it is an existing process. The process design activities such as layout, piping design happen during this phase (Figure 2.6). Construction, if necessary, also occurs during this stage. During Phase III, the automation is designed and implemented. The control system is configured and programmed to perform the automation, as defined in Phase II. Anything necessary to show how the control strategy is implemented is documented. In many control systems, the configuration and programming are self-documenting. Thus, the amount of documentation that needs to be manually created is typically low. The control system to be used must be determined before Phase III can be executed.

Phase III Objectives

- Implement the control strategy defined in Phase II
- Develop operator interface
- Create system qualification and acceptance criteria
- Document anything that is system-specific

Phase III Flow

- Implement control strategy from the bottom up
- Implement procedural control and higher level control

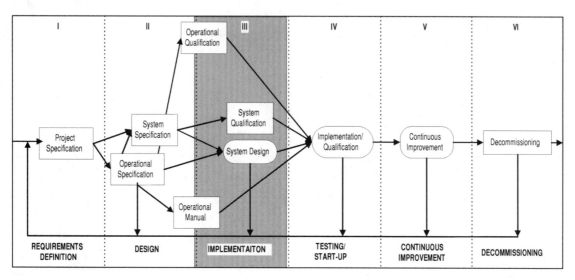

Figure 2.6 Methodology III.

Phase III Deliverables

- Finalized and approved SOPs
- Implemented system based on operational specification
- System qualification and acceptance criteria
- Process simulation (if needed)
- Finalized and approved operational specification

Phase IV—Testing and Start-up

During Phase IV the automation is installed, tested, and the process started up (Figure 2.7). The test plans that were defined in Phases II and III are executed. If this is to be a validated process, the validation is performed during Phase IV.

Phase IV Objectives

- Train operators
- Conduct simulation (if required)
- Test the control strategy defined in Phase II and implemented in Phase III
- Deliver automation strategy that works

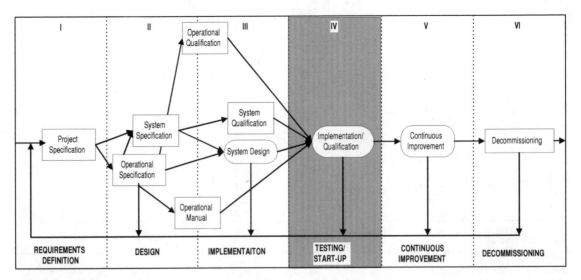

Figure 2.7   Methodology IV.

Phase IV Flow

- Train operators
- Execute qualifications

Phase IV Deliverables

- Working system based on operational specification
- Documentation of operator training
- Documentation of qualification

Phase V—Continuous Improvement

During Phase V the system is operational (Figure 2.8). Any change to the system requires all previous phases to be reevaluated.

Phase V Objectives

- Assure that the system remains operational
- Back up the system

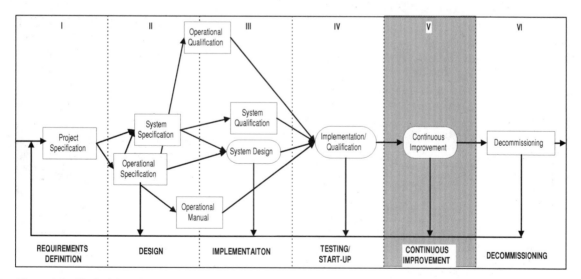

Figure 2.8   Methodology V.

Phase V Flow

- Any change should be evaluated back to Phase I

Phase V Deliverables

- Measurement of automation results
- Benchmarks for other projects

Phase VI—Decommissioning

During Phase VI the system is decommissioned and removed (Figure 2.9). Before this phase is performed, all previous phases must be evaluated for necessary changes. How does this affect upstream and downstream processes? This phase is performed whenever any portion of a process or system is removed. All of the documentation for the system is updated or marked obsolete.

Phase VI Objectives

- Evaluate earlier phases for impact
- Update all documentation
- Evaluate impact on upstream and downstream processes

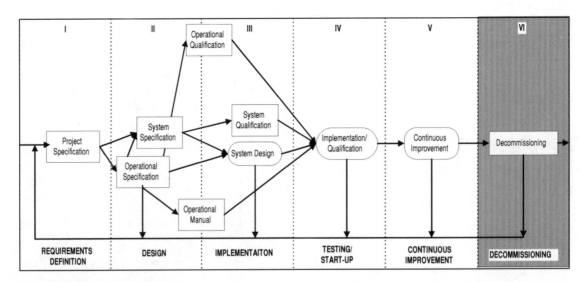

Figure 2.9 Methodology VI.

Phase VI Flow

- Develop plan for removals
- Update documentation
- Remove equipment and automation software

Phase VI Deliverables

- Documentation of the removal

Change Management

Change management is built into the methodology. Changes are not made to the process or system without fully evaluating their impact. This means that whenever anything changes, no matter what phase you are in, you evaluate its impact on all previous phases. The further along in the project, the more expensive the change. This cost of change is the primary reason to front-end-load the definition.

Your change management procedure needs to account for the cost of change (Figure 2.10). Changes early in a project need lower levels of review and approval. The further into a project, the higher the level of

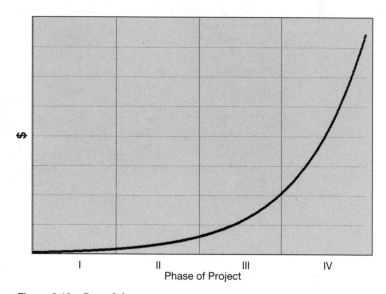

Figure 2.10   Cost of change.

review and approval needed. Changes late in a project can totally ruin the return on investment (ROI) of the project and make the project unsuccessful.

Summary

LARRY: "I can see how this could definitely help manage changes. Changes late in the project are one of our major problems in project management. On the other hand, where is the money going to come from to do this? It seems like this is all going to take a lot of time and money, and our capital budgets are already tight."

VICTOR: "Larry, we may have been very good about delivering projects on budget, but we haven't been getting the ROI out of our automation that we budgeted. We have to realistically define the amount of time that it will take to implement a successful automation project and then base the benefit calculations on those numbers. If the benefits are real, then money will be there to do the project correctly. If not, we shouldn't be messing around with automation in the first place."

DONNA: "If what you are saying about being able to develop the automation in a way that it is independent of the products to be produced in the process is true, then we could see a great savings in product development cost and time to market for new products."

STEVE: "Are you trying to tell me that in Phase II of a project you are going to write the standard operating procedures for the process? That's impossible! You don't even know where the equipment is going to be located yet."

Owen, who has been sitting in the corner silently up to this point says "an experienced operator could define a rough SOP outline for a process long before the equipment location is defined. What does that have to do with automation, though?"

PATTY: "Aren't the SOPs tied in very closely to the operational spec? The SOP defines how the operator will use the control system and equipment to operate a process within defined limits. Sometimes SOPs are defined for nonproduction-related activities that are required to support safe production. Examples of nonproduction procedures may include cleaning, emergency response, and documentation."

ROBERT: "At what point would we start doing things like HAZOPs?"

MARK: "I've been through several HAZOP's and from what I can tell, there is enough information after Phase II to start a HAZOP of the system. You know how the process needs to operate and what your safety envelope is. You can define shutdown criteria and critical alarms when defining the modules in the operational specifications.

In short, you should know what the system should do in hazardous situations for the equipment, such as relief systems, and protective systems. However, you may need all the P&IDs to complete the HAZOP study."

IRENE: "All I can say is that I wish I had had the information we are talking about during my last project. It would have saved a lot of headaches and a lot of guess work on my part. I hate to admit it, but the process would probably also work better."

LARRY: "This is all well and good, but Allen, do you have any numbers for how much time each of these phases will take for a typical project?"

ALLEN: "Historically, we have found that a successful project will spend about 50% of the total time before any coding begins (see Figure 2.11). By doing this, the design and implementation take much less time, and the process performs as desired. Once a project is implemented, the amount of continued support through Phases V and VI varies by process. If a process changes frequently or the control system is large and complicated, annual support can be up to 20 or 30%

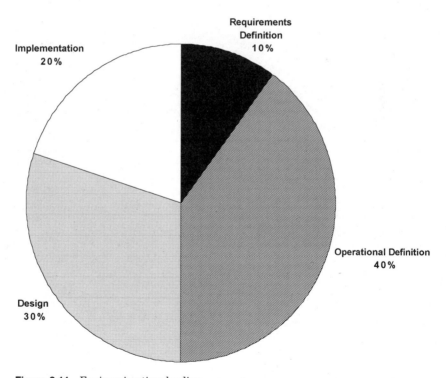

Figure 2.11   Engineering time loading.

of the costs required to initially implement the automation in Phases III and IV. For processes in which the recipes may change but the equipment and automation remain basically static, annual support may be as low as 5 to 10% of the initial costs to implement the automation per year."

VICTOR: "This sounds like it has the potential to have a major impact on both the effectiveness of our automation design process and the benefits derived from our automation investment. Where do we go from here?"

LARRY: "You said before that we were going to have this team implement the next project using this methodology. We can then evaluate it and work out any of the bugs."

ALLEN: "Sounds good. Before we get started though, we should talk about some other issues in automation."

3

Regulatory Issues

ALLEN: "This time we're going to talk about some regulatory issues that have an impact on automation, such as validation."

STEVE: "I recently worked on a start-up of a process that supplies the pharmaceutical industry. Talk about time consuming! We had to validate the process and the control system. IQ, OQ, CQ, PQ, who can keep all of those acronyms straight? It took about 30% of the entire project time just to do the validation. Everything had to be done in black pen and approved by 12 people, I swear. We used more paper during that validation than I've ever seen a project use."

ALLEN: "How did the process work when you were done with the validation?"

STEVE: "Actually, now that you mention it, it was one of the smoothest start-ups I've ever been on. When we were done, everything worked."

ALLEN: "That's what I expected. Validation, although time consuming, can reduce the problems during start-up. It also assures that the process will perform as specified every time."

PATTY: "What do you mean 'the process will perform as specified every time?' Don't processes usually perform as designed?"

OWEN: "I've yet to see a process work right the first time. It takes quite a while for us to figure out how to bypass and override the controls to get it to make product. I know that you've good intentions when designing processes, but you need to get out there to see how they actually run. Allen, what is this validation thing? I think I'd like to hear more about it."

ALLEN: "We'll soon get to validation, but first I want to talk about all of the regulations that touch process automation. Regulation is a law or rule prescribed by a regulating body such as an agency of the government. Most regulations that impact automation are related to safety. The regulatory requirements can be broadly classified into two groups: environmental, health, and safety (EHS); and quality.

EHS Regulations

Environmental, health, and safety regulations apply to all industries, but even more so to the process industry. In the United States, public agencies, such as The Environmental Protection Agency (EPA),The Occupational Safety and Health Administration (OSHA), and The Food and Drug Administration (FDA) enforce these regulations.

EPA's key regulatory requirements govern the release of regulated substances. The EPA grants permits and requires tracking and reporting releases of these substances. The OSHA requirements include safety standards (i.e., protective equipment for personal safety) and prevention regulations. Some of the OSHA regulations that influence the life cycle of an automation system are presented here.

The Emergency Action Plan (29 CFR 1910.38(a))

The Emergency Action Plan standard states that each plant should have an emergency action plan. The plan should have evacuation procedures, safe shutdown protocols for critical equipment, rescue, and first aid information.

Process Safety Management (PSM) (29 CFR 1910.119)

The Process Safety Management standard suggests that in order to prevent process-related incidents and injuries, management systems must be used to identify, understand, and control process hazards. PSM is concerned with preventing or at least minimizing the consequences of catastrophic release of hazardous chemicals. It advocates a comprehensive hazard analysis program to evaluate the process. Furthermore, it recommends that several lines of defense be incorporated in the design and operation of the process. First, well-documented change control policies and procedures, both administrative and operational, should be in place. Second, advance consideration should be given (in areas of safety and the environment) and approvals acquired before any change is implemented to materials, equipment, technology, or utilities.

Given the history of equipment failures that have lead to catastrophes, the regulation also requires that a preventative maintenance (PM) program, related to equipment performance, be in place.

A PM program requires the identification of the critical equipment and devices, establishment of testing frequencies, and documentation of inspections and tests. It also states that appropriate industry standards should be followed.

Several industry-sponsored organizations have developed standards for risk management and risk reduction methods based on engineering principles. ISA has published a standard entitled "Application of Safety Instrumented Systems for the Process Industries, S84.01 (1996)." This standard discusses the design of Safety Instrumented Systems (SIS) that can be a part of a PSM program. Another good source of information on PSM for automation engineers is AIChE's "Guidelines for Safe automation of Chemical Processes (1993)."

Hazardous Waste Operations and Emergency Response (29 CFR 1910.120)

OSHA's Hazardous Waste Operations and Emergency Response requires the establishment of procedures for decontamination, an emergency response plan, and a safety plan. The safety plan should identify hazards and should include briefings prior to any maintenance work.

The Hazard Communication Standard (29 CFR 1910.1200)

The Hazard Communication Standard requires that lists be made of chemicals present on site and Material Safety Data Sheets (MSDS) be accessible for all materials in use (i.e., raw materials, consumables, intermediates, by-products, and products).

Safety Instrumented Systems

Safety Instrumented System (SIS) is a broad definition of an interlocked system. Most processing plants that have a potential for explosion have a separate shutdown system in addition to the process control system.

All companies want to be profitable and be safe at the same time. Safety systems help to manage process exceptions that could be potentially dangerous. Each process industry has unique functional requirements for shutdown systems. For hydrocarbon industries, an emergency could be defined as any process operating outside safety limits. The emergency could be due to equipment failure.

Most of the time, safety systems serve the dual purposes of protection and safety. Protection includes the minimization of risk for assets such as people and equipment. Safety systems may also have fault avoidance as a goal. Today's safety standards do not tell how to implement a safety system.

Managing Risk

IEC-1508 is a draft international standard that outlines the use of electric, electronic, and programmable systems in safety applications in all industries. This seven-part draft is a performance standard. Part 1 of the draft recommends the use of a safety lifecycle. Part 2 identifies methods to qualify and quantify process risk and allocation of the risk to safety systems. Part 3 outlines software requirements and addresses integration issues. The IEC is due to adopt SS84.01 through IEC-1511 soon. Approval agencies may use a base standard for certification of safety systems.

Process Validation

Automation systems increasingly use computers for automating production and quality assurance operations in industries regulated by the FDA. FDA requires that proper installation, operation, and performance of computerized systems be demonstrated and documented before being used in the manufacture of products.

Process validation, as defined by the FDA, is "establishing documented evidence which provides a high degree of assurance that a specific process will consistently produce a product meeting its predetermined specifications and quality attributes." Validation as a concept was proposed in 1976 and was prescribed in "Current Good Manufacturing Practices for Finished Pharmaceuticals, 21 CFR 211" (cGMP) in 1979 (Figure 3.1).

Rule of Thumb for Validation

1. Say what you are going to do.
2. Do what you said you were going to do.
3. Manage all changes.

Produce independently verifiable evidence

Figure 3.1 Validation principles.

The blue book, "Guide to Inspection of Computerized Systems in Drug Processing," was published in 1983 as a guide for its inspectors and is used as a benchmark for systems validation. The following excerpt references validation:

(a) Automatic, mechanical or electronic equipment or other types of equipment, including computers, or related systems that will perform a function satisfactorily, may be used in the manufacture, processing, packing, and holding of a drug product. If such equipment is so used, it shall be routinely calibrated, inspected, or checked according to assure proper performance. (b) Appropriate controls shall be exercised over computer or related systems to assure that changes in master production and control records or other records are instituted only by authorized personnel. Input and output from the computer or related system of formulas or other records or data shall be checked for accuracy. (From "Current Good Manufacturing Practice for Finished Pharmaceuticals" (21 CFR, 211.68))

International Regulations

Good manufacturing practice (GMP) concepts are being widely applied to manufacture of pharmaceutical products around the world. The European Community (EC) is moving toward a unified standard code of regulations for member countries. The EC guide to GMP includes recommendations on environmental air quality and validation.

The UK Pharmaceutical Industry Computer Systems Validation Forum, now called the GAMP Forum, started an effort to promote communication between suppliers and users of automated systems. The forum published draft guidelines entitled "Draft Guidelines on Good Automated Manufacturing Practice" for suppliers of automated systems to the pharmaceutical manufacturing industry. The guide uses a prospective validation life cycle model. This guideline offers good methodology for defining and validating custom solutions and configurable systems. The document identifies several phases for successful validation (see Figure 3.2).

Remember the first rule of validation—"say what you are going to do." This means that you need to have good specifications that are based on standards. The GAMP Forum document recommends formats and a content description outline for all specifications. A well-written specification is the first step to produce a verifiable system.

With increased globalization of the economy, companies will start to rely on consensus standards at the international level. This trend is fairly evident in the pharmaceutical industry. The International Conference on Harmonization (ICH) is working on technical guidelines

- Identify the system
- Produce validation plan
- Produce user requirements specification
- Audit and select supplier
- Review and approve specifications
- Monitor development of system
- Review and approve test specifications
- Produce validation report
- Maintain system

Figure 3.2 Validation life cycle model.

and requirements for drug registration of new products. The goal is more economical use of resources and elimination of delays in approval processes while maintaining the highest product quality, safety, and efficacy. These guidelines will cover GMPs and hence automated systems.

Quality

ISO 9000 ensures that a certified company has procedures in place and follows those procedures to guarantee consistent product quality. There is increasing demand from consumers that suppliers have quality certification. Remember, ISO 9000 is an accreditation system.

How Automation Systems Can Help

These regulations require enormous amounts of information to be managed at the source of the process. The automation methodology outlined in later sections helps you to comply with these regulations and gain competitive advantage while doing so. All of the regulations require a high degree of documentation. The methodology is based on a detailed and structured definition of how automation operates the process. Since everything is defined on a modular level, checkouts are performed on modules first and then on interactions between modules. This greatly simplifies validation. The automation methodology has validation built into it. A successful approach to validation will be presented in Section 3. Change control is an important aspect of all of the regulations. The modular nature of the system allows changes to be implemented and tested on a modular level. This greatly simplifies the implementation and testing of changes.

Summary

LARRY: "All of these regulations seem to just add a lot of overhead."

ALLEN: "Just remember, regulatory agencies, especially the FDA, assume that you haven't done it right unless you've documented what you've done, how you've done it, and what the results were. There are two ways that you can approach regulations. You can react to them, or approach them in a proactive manner. By applying good engineering practices and a modular approach to automation, you're well on your way to complying with most regulations. Since we're on the subject of good engineering practices, let's next discuss one of the industry standards that can help, S88.01."

4

S88 Models and Terminology

ALLEN: "Since we're working through a project, can someone tell me about the process?"

PATTY: "We're building a flexible, multiproduct facility with recipe-driven execution. We're ready to start defining the process. Larry and I are ready to start the process design."

VICTOR: "Allen, you mentioned a standard for doing batch manufacturing. Can you help us understand what this standard is all about? Is it relevant to what we are trying to accomplish?"

ALLEN: "Certainly." [Allen's description of the S88 standard follows.]

S88 is an international standard that defines terminology and models for batch control. The standard can also help in the design of batch processes by characterizing and defining them. The standard provides common terminology but not interoperability among different systems. However, one can use the terminology and models as a consistent means of managing recipes over multiple sites. S88 as a standard can help to emphasize good practices not only for design, but for operation of batch manufacturing plants.

The S88 Part 1 standard is the result of the first part of this two-part initiative to address batch control. It is called "ISA -S88.01, Batch Control, Part 1: Models and Terminology." The SP88 committee comprises process manufacturers (users) and control vendors from North America and Europe. This standard helps to establish a universal model for batch control. A universal model helps both users and vendors use a common terminology. This reduces the potential for misunderstandings in communication by providing a common language or notation. For example, in the past the term "recipe" meant different things to different people. The standard defines many terms, such as "recipe," so that they can be commonly understood.

A batch process is difficult to describe in words. Users were looking for a consistent way to define the operational and functional aspects of a batch process. This standard helps to present this collective knowledge in a simple and elegant way. The automation systems that enable the production may come and go over the life cycle of a process.

S88 Background

The International Purdue Workshop on Industrial Computer Systems, Special Applications Programming Committee (TC-4) worked on a design toolkit in the early eighties that provided standard terminology for batch control.

A status paper was issued by the Working Group 6 of NAMUR in 1987. This status paper presented an approach of defining recipes, basic operation concepts, procedure blocks, and batch activity models. NAMUR (Normenarbeitsgemeninschaft fur Meb- und Regelungstechnik in der Chemischen Industrie" or "Work Association for Measure and Control technology in the Chemical Industry") has been active since 1949 and generates recommended practices and guidelines. The SP88 committee drew heavily on this work in the initial stages.

The S88.01 defines a set of models and terminology for

- Batch processes and equipment
- Batch control concepts
- Batch control activities and functions

A batch process can be defined as a process that leads to the production of finite quantities of materials by subjecting quantities of input materials to an ordered set of processing activities over a finite period of time, using one or more pieces of equipment.

Batch Definition

S88.01 defines a batch as:

1. The material that is being produced or that has been produced by a single execution of a batch process.
2. An entity that represents the production of a material at any point in the process.

NOTE: Batch means both the material made by and during the process and also an entity that represents the production of that material. Batch is used as an abstract contraction of the words "the production of a batch."

Recipe Independence of Equipment

The recipe is a data structure that contains a header (production information), formula, procedure, equipment requirements, and other information. S88.01 indicates that a recipe should be independent of the equipment on which it will be executed. This separation of equipment and procedure lends itself to procedural elements that are transportable between common pieces of equipment.

The standard describes four types of recipes (discussed later). A recipe is an entity that contains a minimum set of information that uniquely defines the manufacturing requirements for a specific product.

Process Classification

The standard classifies the process based on the number of products it makes (recipes) and the physical connections (process). The process can be single-path, multiple-path (with isolated parallel paths), or networked (with fixed or variable connections) (Figure 4.1).

In a single-path process, the batch passes sequentially in a predefined order from one unit to the next. In a multiple-path process, several batches may be active at a time. The equipment may be dissimilar in physical characteristics. For example, a reaction operation may be handled by two units in one path and by just one unit on the other path.

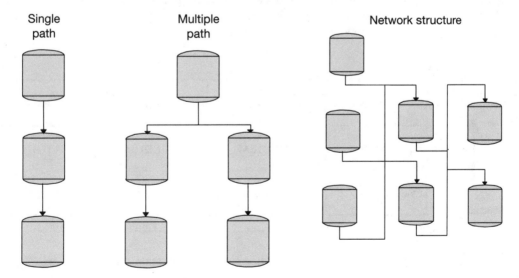

Figure 4.1 Process structure.

In a network structure, the sequence of the units may be determined just prior to or during the execution of the batch or be preassigned.

The appropriate path is determined at the moment of execution based on the constraints, such as equipment capability and recipe requirements. A network process is the most complex to control. This is due to the need for allocation of equipment and the arbitration of requests for that equipment. Figure 4.2 shows the complexity of batch processes based on number of products and process structure. The more complex the process, the more need there is for allocation and arbitration, and, therefore, a batch management solution. The more products that will be produced in the manufacturing process, the more need there is for recipe management, and, therefore, a batch management solution.

A continuous process can be defined using this terminology as well. In a continuous process, the complexity of the control functionality relating to procedural control is minimized. Typically, continuous control can be achieved with equipment entities alone. The models in the standard are:

- Process model: a model for subdivision of a batch process
- Physical model: a model that can be used to describe the physical assets of the enterprise
- Procedural control model: a model for the subdivision of procedural elements that can be typically found in batch processes
- Procedural control/equipment mapping model: a model that defines the links between the three previously mentioned models to achieve process functionality

Many Products	**HIGH**	**HIGH**	**HIGH**
Few Products	**MEDIUM**	**MEDIUM**	**HIGH**
One Product	**LOW**	**LOW**	**MEDIUM**
	Single Path	Multiple Path	Network

Recipe Management ↑

Allocation and Arbitration ⟶

Figure 4.2 Batch applicability.

- Recipe types model: a model that defines the recipe types and relationships between the different recipe types
- Control activity model: a hierarchical model that deals with the functional activities of a batch system

Process Model

A batch process can be subdivided and organized in a hierarchical fashion. This is referred to as the process model (Figure 4.3).

Process

A process is a sequence, or defined order, of chemical, physical, or biological activities for conversion, transport, or storage of material or energy. Any process that has a finite quantity of material output is classified as a batch process.

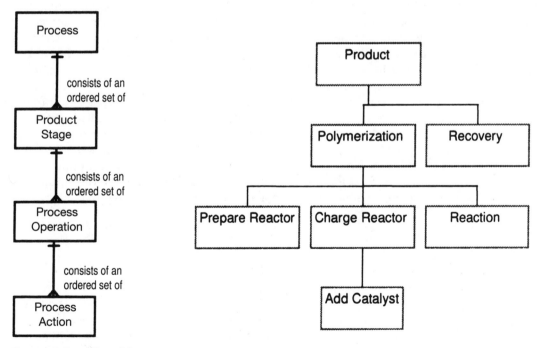

Figure 4.3 Process model.

Process Stage

A process stage describes a major function that usually results in a planned sequence of chemical and/or physical changes in the material being processed (e.g., polymerization). A process stage operates independently from other process stages.

Process Operation

A process operation represents a major processing activity. An operation usually results in a chemical or physical change in the material being handled (e.g., prepare reactor, charge, or react).

Process Actions

Each process operation can be subdivided into an ordered set of one or more process actions. The process action describes a minor processing activity that is required to make a process operation (e.g., add catalyst, heat reactor).

The Physical Model

The physical model is used to describe the physical assets of an enterprise. The physical assets are usually organized in a hierarchical fashion. The model has seven levels starting at "enterprise." The first three levels are not described in detail in the standard (Figure 4.4).

The lower four levels are specific to equipment types. Lower-level groupings are combined to form higher levels in the hierarchy. In some cases, a grouping within one level may be incorporated into another grouping at the same level. These levels are typically defined by engineering activities. During these engineering activities, the equipment at lower levels is grouped together to form a new higher-level grouping. This is done to simplify operation of that equipment by treating it as a large piece of equipment. Once created, the equipment cannot be separated except by reengineering the equipment at that level.

Process Cell

A process cell is the span of logical control of one set of process equipment within an area that can be used to manufacture batch product. Note, though, that a product in this context may not be a final product for sale. It may be an intermediate product to be used in other processes. The existence of a process cell allows for production scheduling on a process cell basis. This also helps in design of control functions and actions. The domain for the batch control system is the

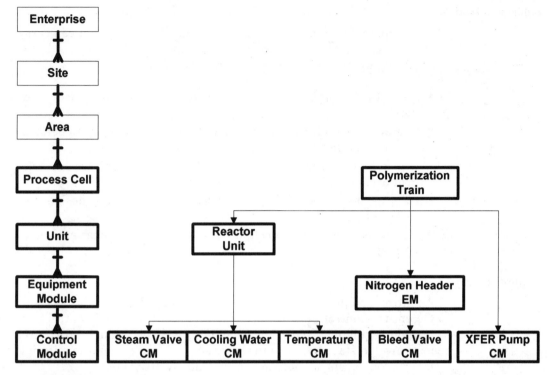

Figure 4.4 Physical model.

process cell. In many cases it is a combination of systems supplied by more than one vendor and manual procedural controls.

Unit

A unit is a collection of equipment modules and control modules. The modules may be contained within the unit or acquired temporarily to carry out a major processing activity. A unit may have any of several key attributes:

- Contains a flexible amount of equipment, equipment modules, and control modules
- Operates on all or part of the batch
- Operates on only one batch at a time
- Cannot acquire another unit
- Operates independently of other units

Equipment Module

An equipment module is a functional group of modules that can carry out a finite number of specific minor processing activities (i.e., a header). An equipment module has several key attributes:

- Consists of equipment, control modules, and other equipment modules
- Contains all equipment and control functions necessary to perform its process function
- Is usually centered around a fixed piece of equipment; for example, heat exchanger, weigh tank, agitator
- Its scope is defined by the finite processing activity it is designed to carry out
- May be part of a process cell, unit, or another equipment module

Control Module

A control module is the lowest level of grouping that operates as a single entity. Control modules perform basic control. Basic control may include regulatory control or sequential control. Sequential control may involve state-based control. Exception handling and monitoring is included. A control module has several key attributes:

- It is typically a collection of sensors, actuators, and other control modules
- It may be made up of another control module(s) (e.g., a cascaded loop)

The Procedural Control Model

The procedural control model describes control that directs equipment-oriented actions to take place in an ordered sequence to carry out some process-oriented task. These procedural elements are organized into a hierarchy (see Figure 4.5).

Procedure

Procedure is the strategy for carrying out a process. The domain of the procedure is the process cell. The procedure is the highest level in the hierarchy. It is defined in terms of an ordered set of nit procedures.

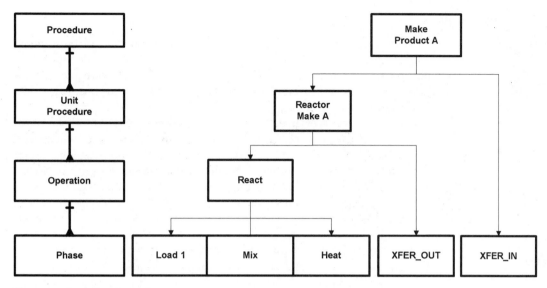

Figure 4.5 Procedural model.

Unit Procedure

A unit procedure is a strategy for carrying out a contiguous process within a single unit. It consists of Operations and methods required for initiation, organization, and control. Multiple-unit procedures of one procedure may run concurrently, each in different units.

Operation

An operation is an ordered set of phases. It will usually take the material being processed from one state to another, involving chemical, physical, or biological change. According to the standard, only one operation is presumed to be active on a unit at a given time. An operation is carried to completion within a single unit.

Phase

A phase is the smallest element of the procedural control model. It accomplishes a specific process-oriented task. Phases can be executed in a parallel or sequential manner. Phases can be self-terminating, that is, end without being commanded to do so. A phase is the lowest group of process actions. The design of a phase needs to take into account safety and exception conditions.

Equipment Entities

An equipment entity is formed from the combination of equipment control (brain) and physical equipment (brawn). Equipment control is equipment-specific functionality that provides the actual control capability and is classified into three types of control: procedural, basic, and coordination control.

Procedural control directs equipment-oriented actions to take place in an ordered sequence in order to carry out some process-oriented task. Basic control is dedicated to establishing and maintaining a specific state or process condition. Basic control includes regulatory control, interlocking, monitoring, exception handling, and sequential control. Coordination control directs, initiates, and/or modifies the execution of procedural control and the utilization of equipment entities.

A functional distinction exists between different types of control. Procedural control consists of procedural elements. We identified the hierarchy of procedural elements in the procedural control model.

A phase is the lowest element in the procedural hierarchy. The execution of a phase may result in commands to basic control or to other phases.

Basic control is similar to control of continuous processes. It performs the key functions of monitoring, controlling, annunciating, and interlocking. For batch processes, there may be additional commands that need to be acted upon.

Coordination control is a type of control that directs, initiates, and/or modifies the execution of procedural control and the utilization of equipment entities. Some examples of coordination control include propagating modes, allocating resources, and arbitrating common resource equipment requests.

The relationship between the procedural control model, the physical model, and the process model is depicted in Figure 4.6. Procedural control may be defined as a part of equipment control or it may be passed on via a procedural entity from the recipe.

Recipes

The standard defines four types of recipes, namely general, site, master, and control (Figure 4.7). Within an enterprise, a general recipe might describe the requirements to make a product to various sites. A site recipe will describe specific requirements in terms specific to that site. Master recipes are derived from a general or site recipe and account for specific information (e.g., equipment arrangements) that are

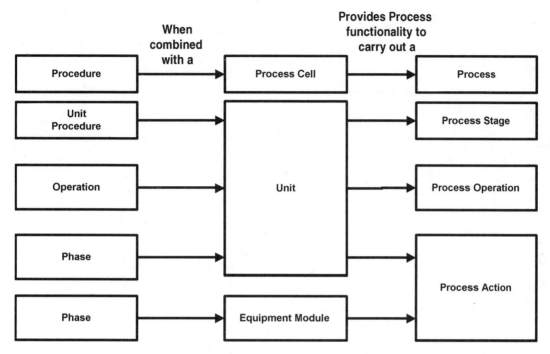

Figure 4.6 Model integration.

necessary to make the product. Last, a control recipe starts as a copy of a master recipe. The control recipe is executed to make a batch of product. The recipe levels are collapsible. For example, a master recipe can be created as a standalone entity without the need for a general recipe or a site recipe. However, this assumes that the recipe developer has full knowledge of the specific process and its capabilities.

Recipe Procedure versus Equipment Procedure

At some point, a recipe procedural element must map to an equipment procedural element. This means that a recipe procedural element points to or calls equipment procedural elements. In most cases, this mapping is accomplished at the phase level. Typically, a batch system will have a recipe phase that references an equipment phase (Figure 4.8).

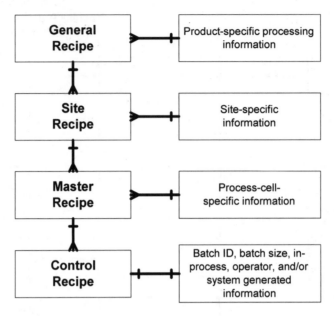

Figure 4.7 Recipe hierarchy.

Control Activity Model

The control activity model is functional model that helps describe the functional requirements of a batch control solution in a succinct way (Figure 4.9). This will help in selecting the right batch solution. The standard does not design a solution, nor is it implementation specific. In other words, any applied vendor technology could be used to implement the batch control functional requirements.

Summary

VICTOR: "Can all of this help to reduce engineering costs?"

IRENE: "You can model the process and define the phases before you buy a single piece of equipment."

PATTY: "I like the fact that the recipe procedures can be changed to make different products. We'll no longer need a programmer to build our recipes."

ALLEN: "The standard can also help in reducing the installation costs. If the standard is used as a framework for communications for

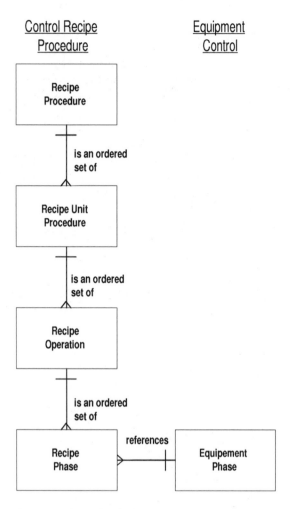

Figure 4.8 Procedural mapping.

the whole project team, then the design and implementation costs will be less."

IRENE: "What about our existing plants?"

ALLEN: "The standard does not suggest that you abandon the current way of dealing with batch. The terminology will provide a reference to talk about your existing process."

VICTOR: "Sounds like we need to be applying this to our automation projects. Are there any caveats?"

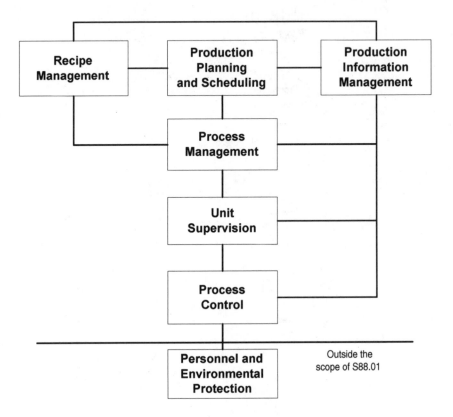

Figure 4.9 Control activity model.

ALLEN: "The standard is not a compliance standard. It is up to you to implement the standard. Carefully look at your manufacturing needs before deciding on what levels of models need to be implemented."

5

Know Your Automation Objective

PATTY: "That sure was a lot of information! Standards are good, but how do we apply the standards to our projects?"

VICTOR: "Earlier we talked about the six phases of a project. It seems that since developing an automation objective is the first step in a project, that is where we should start."

ALLEN: "I couldn't agree more. Let's follow the methodology in our discussions and work through this project. First let's start by defining what we mean by "automation objective." Any ideas?"

LARRY: "Well, according to Webster, an objective is "an aim or end of action'."

ALLEN: "OK. So how does that apply to automation?"

VICTOR: "From a management perspective, objectives help to assure that people know what they should be working on to achieve corporate goals. It seems that the same is true for a project. It gets everyone reading from the same sheet of music and, even more importantly, makes sure that they are reading from the *right* sheet of music."

ALLEN: "So if that is the reason to have an objective, what are some of the possible objectives that you could have for an automation project?"

OWEN: [Said under his breath, not intending for anyone else to hear.] "Put operators out of work."

Steve had worked with Owen on several projects. He heard what Owen said, but didn't want to embarrass him, so he said, "I think that it is to enable operators to continuously do a better job. I think automation could help operators do what we expect them to do anyway—"more with less'."

PATTY: "I think that automation can help us to better understand the process itself. If we can better understand the process, we can improve it even more. I also think that automation can help us understand where the process variability is."

ROBERT: "Patty, are you saying that we don't really understand the processes we are designing?"

PATTY: "We understand what we designed. Over time, though, the process changes, and it isn't always operated in the way we designed it to operate."

STEVE: "How can you expect it to operate as you designed it to operate when you don't tell anyone who is going to operate it how you designed it to operate?"

PATTY: "I guess I always thought that manufacturing would know how it should operate."

ALLEN: "Or maybe you didn't have the tools to communicate the operational requirements. Let's try to avoid assigning blame. Let's just brainstorm and make only positive additions to the list of potential benefits of Automation. Donna, what benefits do you think that automation could have on your job?"

DONNA: "As I said last week, I think that if we could truly design the automation to be independent of product, we could definitely shorten the time to market for new products."

IRENE: "Maybe I'm missing something, but I don't see how you can design the automation independent of the products that the process is going to produce."

ALLEN: "Good point! Does anyone have any ideas?"

MARK: "I guess, if I really think about it, the mechanical process isn't really designed based on the products. It is designed based on processing requirements. Quite often, I don't even know much about what products are going to be made in the process. Is there any way that the automation could be designed the same way?"

ALLEN: "There is, but let's hold off discussing how until we're done with our automation objective, OK?"

LARRY: "Sounds good to me. I think that if what you said last week is true about the methodology, we could have much better project control. It's scary, as a Project Manager, to not understand what the Control Engineer is doing. If we could have a clear specification of what the automation is to do, I could better manage that part of the project. Through better project control, we could assure that projects are kept within budget and on schedule."

VICTOR: "And, of course, deliver the expected results, right Larry?"

LARRY: "Of course."

STEVE: "I've never seen this, but I'd love to see automation help us to see smoother start-ups. If we truly understood the operational requirements and had a well-documented automation strategy, I think we could reduce the number of changes during start-up and reduce a lot of my headaches, but. . . ."

ROBERT: "One of the major benefits that I have already seen from automation is process safety. Interlocks, emergency shut-down, and alarms help to keep processes from unsafe conditions."

ALLEN: "Owen, you've been kind of quiet. What do you think can be achieved with automation?"

OWEN: "Over the years, I've seen a lot of lofty ideas on what people were going to do with automation, but I've yet to see the real benefit."

ALLEN: "OK. If there was one thing that automation could do to help you to better do your job, what would it be?"

OWEN: "I've seen one problem that automation has caused that it might be able to fix. Get rid of all of the alarms. I know that Robert says that alarms have helped to make processes safer, but they just give me a headache."

VICTOR: "Guys, what I'd like to see is that this stuff is paying for itself. I believe that it can, but we need to start showing concrete benefits—things that I can take to the board to justify the investment."

ALLEN: "Excellent point Victor. We've been talking about a lot of things from the process and plant perspectives. What about the business drivers? How can automation help to achieve measurable benefits that the business managers will see and understand?"

LARRY: "Our automation objective needs to be measurable, specific, and reachable. I've seen too many objectives that were meaningless because they were unrealistic, too broad, or too 'motherhood and apple pie'."

STEVE: "I don't know about anyone else, but I agree with Larry. It sometimes feels like we're applying technology for the sake of technology; 'techno-lust' so to speak."

ROBERT: "Steve, that is exactly why we need to define the 'automation objective.' That will keep us from doing exactly what you said. There has to be a reason for doing what we do. Something that I think we could achieve through automation is better information about our processes and products. All too often we have either too little, too much, or the wrong information. We need to define our information requirements both from a process and product perspective. I definitely don't mean more information. I mean better information."

ALLEN: "So in other words, you want the process to 'communicate' with you rather than to simply store information?"

ROBERT: "Exactly!"

ALLEN: "There must be a business driver behind automating a process. If there is not a clear business reason for the project defined before you start, then there will be little, if any, benefit realized."

Business Drivers

Businesses pursue goals that are tracked based on traditional measures of financial performance (profitability, market position, growth, and risk), or, intangibles such as contribution to society and economic security for its employees. Some of those things are internal to the business and some are external, depending on the challenges and environment. The business drivers are methods that businesses pursue to meet these goals. These goals are further defined in the hierarchial structure of the organization including functional areas. The functional areas such as R&D, supply (manufacturing, distribution), and marketing may interpret the business goals and draw their own strategies. The strategies are usually consistent across functional areas. Manufacturing strategy primarily drives automation strategy. Elements of the automation strategy are driven by business goals.

The following is a sampling of some common business drivers:

- Competition—open markets
- Rapid commercialization
- Increasing customer demands
- Increasing regulation (ISO, OSHA, FDA) verification
- Increasing process complexity
- Expecting operators to do more with less
- Increasing product variety
- Decreasing time to market

Automation Objectives

There are many automation objectives that can be defined to meet the business drivers. There will typically be several automation objectives for a given process. For example, the business driver for a project may be to delay building a new plant by increasing the output of an existing plant by 20%. From that, the automation objective for the process might be to increase overall process capacity by 20%, through increased operating efficiency and decreased turnaround time.

The following is a list of some common automation objectives:

- increase process flexibility
- improve process repeatability
- decrease product cost
- improve product quality
- improve process reliability

- ensure process safety
- increase process capacity
- better understanding of the process
- process design upgrade
- improve operating efficiency
- improve job content and satisfaction
- recipe improvement
- streamline manufacturing environment
- improve equipment availability
- improve management information
- capture knowledge

Notice that reduction in workforce is not on the list. It is unlikely that automation will reduce the number of people that you need to employ. It will, however, make your employees more productive. Although you may need less people operating the process, you will need more people supporting the automation and information systems. It will shift the skills that are required within your workforce.

Flexibility versus Repeatability

There are two fundamental ways in which automation can improve your process; it can add repeatability and it can add flexibility. In most cases, you either want a given part of the process (module) to run exactly the same way every time (e.g., a compressor) or you want to be able to determine how it will operate at run time (e.g., a batch reactor). Your job is to define the requirements for each part of the process. You can then combine these flexible and/or repeatable modules to produce the efficiency, quality, product flexibility, reliability, safety, or capacity utilization levels you outlined in your strategy.

When considering an entire process, there are few cases in which total flexibility with little repeatability is needed, and vice versa. In batch manufacturing environments, regulations (among other things) are driving the need for proven repeatability. In continuous processing environments, customer demands and other factors are driving the need for flexibility. The need for modularity and definition are the same in either case. The differences show up in the number of changeable parameters and the order of operations. In a more repeatable system, the parameters and order of operations are fixed, whereas in a flexible system the parameters and order of operations are both changeable (Figure 5.1).

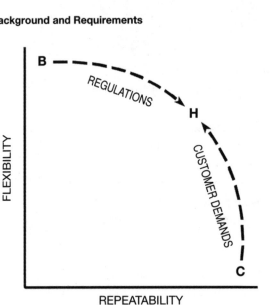

B: Totally Batch Process
C: Totally Continuous Process
H: Hybrid Process

Figure 5.1. Flexibility versus repeatability.

External forces are changing the requirements in many processing environments. Many historically continuous processing environments are facing pressures to become more flexible. Customers are demanding individual specifications for what were once known as commodities. If you don't proactively begin to deliver what the customer wants, your competitors will. The discrete manufacturing industries saw this trend several years ago. They responded by manufacturing lot sizes of one part. Modular Batch Automation is the process industry's response to the same external pressures.

On the other end of the spectrum, historically batch processing environments (e.g., food, beverage, pharmaceutical) are facing increasing regulations. These regulations are forcing them to prove repeatability in their processes. ISO, FDA, OSHA, and other standard and regulatory agencies want to know that the process is operating exactly as designed. Quality product must be produced in a safe manner every time.

Flexibility

When automating a process or module of a process for flexibility it is important to remember this concept: *Do not impose limitations on the process through automation that are not physical process limitations.* In other words, make sure that the Automation is capable of doing anything that the process is capable of doing. Assume that any artificial limitation imposed by the automation will need to be removed at some time in the future.

Define the automation based on the capabilities of the process, not based on the products that will be produced by it. The automation should then be reviewed to assure that the product recipes will be able to run. For example, do not enforce a purge step for a reactor if there isn't a physical process constraint. If the process would be capable of producing a product that would not require a purge, make the purge a recipe option rather than a required step.

Once the automation is defined, take some product recipes and validate them against the automation. Verify that all of the capabilities necessary to produce the products are available. This will reduce the "gotchas" after coding is complete.

Repeatability

Industry is awash with the concepts of "batch automation," yet a large percentage of processing, even in a batch environment, requires automation principles that are more similar to a continuous process.

Whenever the success of automation revolves around doing it the same way (the best way) every time, the automation is the same as if it were a continuous process. This is true for anything from a feed system, to a compressor, to a fluidized bed, to a solids transfer system. Each of these modules need to operate consistently, repeatably, and with a known response.

Even though critical success factors differ between flexibility and repeatability, the methodology is very similar. Some of the primary differences will be in how fully the operation is defined at the automation level, and how much the automation constrains the process beyond the physical constraints of the process. We will discuss these differences more fully when we talk about procedural control.

LARRY: "I don't think you can define automation objectives from thin air. It sounds to me that we need to have a lot of people involved at this point in the project so that the automation objective is clear. Is that realistic?"

VICTOR: "I don't think that anyone is saying that all of these people need to be involved long-term. Just that they need to be consulted to make sure that their expectations are understood. Larry, is it realistic to do a project that doesn't meet expectations?"

LARRY: "I just want to make sure that everyone understands that this will all take more time and probably cost more money."

IRENE: "I don't know that I agree that it will take more time or cost more money. During my last project, I spent a lot of time redoing work because I didn't know those expectations at the beginning. I think it will take much less programming time. I do agree that it will take more of other people's time, though. But, as Victor said, we will then meet everyone's expectations."

VICTOR: "I've got a question. How do we currently decide who gets involved in a project and when? Irene, when do you usually get involved in a project?"

IRENE: "Actually, I've only been on one project, but I didn't get involved until the process was almost fully designed. The P&IDs were nearly complete when I was assigned to specify the instrumentation. I didn't even know that I was going to be working with the control system until some time later."

VICTOR: "Larry, is that usually when the control engineer is assigned to the project?"

LARRY: " Well, yes."

VICTOR: "So Allen, how do we define who needs to be involved in a project and when they should be involved?"

ALLEN: "All key disciplines should be involved in a project at a very early stage. That doesn't mean that everyone needs to be committed full time to a project, just that they are available to attend some or all of the team meetings. Here is a list of some of the disciplines that should be represented:

- Process control
- Instrumentation
- Manufacturing
- Operations (operator)
- Safety
- Environmental
- Quality assurance
- Regulatory compliance
- Maintenance
- Product development
- Process engineering
- Systems support

ALLEN: "Owen, have you ever been involved during the conceptual design stage of a process?"

OWEN: "Are you kidding?"

ALLEN: "OK. Has anyone ever been involved in a project where an operator was involved early on?"

LARRY: "We can't afford to pull our operators out of manufacturing for every project. They have work to do, and they are overworked as it is."

PATTY: "So instead, we can afford to have our processes not operate as designed?"

LARRY: "OK, OK, so Allen, how much time and money is all of this going to cost?"

ALLEN: "There is a huge range in savings reported by using a modular (S88 aware) approach to automation. The numbers go from a low of 20 to 30% to a high of 70%. The 70% is achieved when duplicating identical equipment. I typically use 20 to 30% as a conservative estimate.

Larry, as you've mentioned, the savings will not be realized on the first project. The first project may actually take more time and money. The true savings will be realized on subsequent projects, and the maintenance of the modular systems. However, the true benefits of a modular approach to automation are typically not in the engineering time or installation cost. The benefits are derived from a flexible and maintainable automation strategy that is in line with the process objective and the business drivers.

So what are the business drivers for this project?"

PATTY: "Here is a block diagram for the process. This is the starting point for the PFDs."

Example

PATTY: "The process consists of 13 distinct physical equipment stations (Figure 5.2). Based on what you said about S88, I doubt in the end that they will all be called units. More likely, there will be a few units and many equipment modules. The heart of the process is the two flexible reactor units. The feeds to the reactors are several raw materials and a catalyst. A few utilities, including an oxygen analyzer and a vacuum system, service the two reactors. Functionally, the reactors need to be able to agitate, heat, cool, react, strip, transfer, purge, blanket, wash, and rinse. Many of these things need to be done at the same time.

Upstream of the reactors are the catalyst and three different raw materials. At this point, assume RAWMAT1 will be a tank loaded from a tanker and both reactors can load from this tank at the same time. RAWMAT2 will be a bag-loading station used by only one reactor at a time. RAWMAT3 will be a plant header supplying material

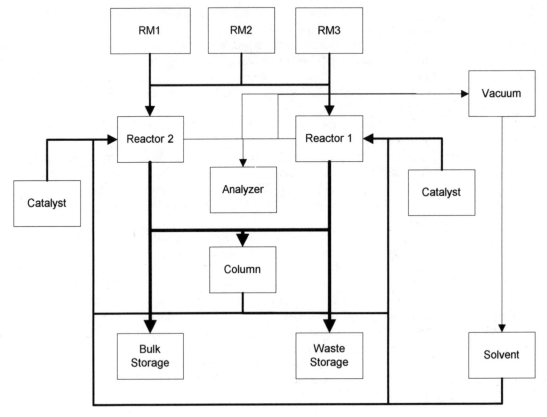

Figure 5.2. Sample process—starting point.

whenever either reactor needs it. Each of the catalyst feeds is a drum-loading station servicing only one reactor.

Downstream of the reactors is a batch distillation column. Only one reactor can feed the distillation column at a time. The column will be designed flexibly enough to allow either the bottoms or the overheads to be sent to bulk storage and the other sent to waste.

Bulk storage has some flexibility of its own. It will allow the vessels to be washed with solvent, apply recipe values to determine which tank receives the product, or bypass the product into waiting tankers.

The thought behind the entire process is to allow for absolute maximum flexibility. We need to allow one reactor to run while the other is being cleaned. If the same product is being campaigned out of one reactor, the other can run but not have the use of the column. Or the process could run both reactors directly to the bulk storage unit into different vessels."

ALLEN: "That is a very good starting point for the process description. What are the business, drivers though?"

PATTY: "In my opinion, the business driver is that this process needs to be able to produce multiple products or have flexibility. We have ten new products slated for this process in the next nine months."

IRENE: "Isn't that one of the automation objectives?"

VICTOR: "I think you're right, Irene. Our business driver is to meet customer demands for product with custom specifications. There is also a market for contract manufacturing that can take up any excess capacity that we will have in this process. That means, as Patty said, we need to be able to produce almost anything in this process. We have little idea what products will be made in this process even next year."

LARRY: "Based on our methodology, what we need is to define these things in measurable, specific, and achievable terms. We then need to define the metrics against which they will be measured."

Automation Objectives for Example Process

The automation objectives for the example process are:

1. Flexibility
   a) Able to make any product that the process is capable of making
   b) Able to take a product from scale-up to production in two days
2. Regulatory compliance
   a) Validated
   b) Safe
3. Detailed batch report produced with each batch of product

Summary

PATTY: "Now that we know the automation objective for the process, Irene should be able to start on the coding."

ALLEN: "Not so fast, Patty. Irene, do you have enough information to start on the coding?"

IRENE: "Not even close. I don't know anything about how the process should operate. At this point, all we've done is define the objectives of the process based on the business drivers for the project. That doesn't tell me what you want the reactors to do."

PATTY: "What do you mean? We said that the reactors each need to agitate, heat, cool, react, strip, transfer, purge, blanket, wash, and rinse. We also said that the process needs to be flexible. What else do you need?"

IRENE: "What do you mean by react? Does agitate have a variable speed agitator and how long do we agitate? There is so much more information that I need that I don't even know where to start."

ALLEN: "Irene is right. We have only begun to scratch the surface of what needs to be defined. I think that this is a good time to start talking about some of those things, and how to go about it."

Operational Definition

6

Define! Define! Define!

ALLEN: "OK. Now it's time to get down to business. Who can tell us the next phase of a project after the requirements definition?"

LARRY: "If we are following the strategic automation methodology, then it would be operational definition. I still don't understand how we're going to define how the process is going to operate this early, though."

PATTY: "I know how the process should operate, the problem is finding a way to communicate that to everyone else."

IRENE: "I definitely agree that we haven't found good ways to communicate that knowledge. The P&IDs were all I got on the last project. What about flowcharts? I drew flowcharts on the last project."

STEVE: "I worked with your flowcharts on the last project, and I've been on other projects where flowcharts were used to document the control code. They're all fine and good if you don't make any changes. As soon as any changes are made, though, they fall apart. People just don't update them, because they are not a convenient tool. Not everyone can look at a flowchart and relate it to what the process is doing. I remember using flowcharts in programming class in college, but they don't seem to work very well any more."

ALLEN: "At one time, flowcharts were heavily used in the software industry. With object oriented-technologies, though, they aren't used often anymore. They have been replaced with things such as entity relationship diagrams. We will talk more about object oriented technology later."

How Should Your Process Operate?

If any one thing could be considered most critical to the success of an Automation effort, an up-front definition of the operational requirements would have to be it. The easiest way to say this it is "How do you know if you are successful if you don't know what you want in the first place?" Many excuses are used to justify why there is no clear definition up front. The most common, "We don't have time," is the least acceptable. What is really being said is "We don't have time to define how our process will operate, so we'll accept that it will take longer to get it running, and we'll even accept that it will not operate as we want it to operate when it is done."

Let us revisit a table that I showed you during our first meeting (Table 1.1, page 12). Many of the tools needed to achieve these advantages weren't available until relatively recently. For instance, although relational databases have been around for quite a while, it's been only in the past few years that they have been capable of storing heterogeneous data such as drawings, word processing documents, video, etc. The primary reason that your definition should be database driven is to eliminate redundancy of data.

Database Definition Tool

Each piece of data should be stored and maintained in one and only one place. In conventional engineering practice, the same information is repeated in multiple locations, and maintained separately by different people and systems. When it comes time to update, the information is not updated in all locations. This creates discrepancies that are not easily reconciled. Most regulations require that information in change control and document management be verifiable and *must* match reality. This means that discrepancies are unacceptable.

By using the database tools that are available today, data integrity can be maintained and redundancies can be minimized. Databases can be used to document the design and also document the implementation for automation. The information can be available to communicate with the operator as to how a particular phase should be running, or what a particular control loop was designed to do. This information also has to be available to the person who is troubleshooting the automation at 2 AM when something doesn't work. If everyone doesn't know what the automation was supposed to do, the automation will receive the blame, even though nine times out of ten, failures are the result of something other than the automation.

We said previously that the focal point of classical control was the loop. This was because nothing above the loop was defined. Automation defines everything up through the interactions within the entire process. This allows each person to choose the level of granularity that they need. What do we mean by granularity? A plant manager can look at an entire process and know what is being made and how the process is being used, whereas an operator can tunnel down and see the status of each valve and motor in the process, if needed.

Automation is defined in terms of process capabilities and requirements. We only look at the products that are to be made with the process to define those requirements. Once the processing requirements are defined, the automation is then defined in terms of equipment and process. This definition is product independent. Later, we will discuss further about how to go about subdividing the process so that this definition makes sense.

Process Definition

Drawings did a fine job of defining the process operation when classical control was used. A control strategy that is based on a P&ID drawing is only a time slice of the process operation, usually steady-state. What happens during start-up, shutdown, and during exceptions? That doesn't even get to the needs of a flexible batch plant. The days of a static picture of a process describing everything you need to know are gone! In order to achieve the level of definition that is required for a successful automation effort, the process must be modularized. Modularization gives the process the flexibility and capability that is required by the automation objective. How to modularize a process will be discussed in Chapter 7.

As Irene indicated, the person making most of the decisions about how a process will operate has traditionally been the engineer programming the control system. If that person has the knowledge of the process designer, the operator, the product developer, the manufacturing engineer, etc., then this is a perfect fit. Otherwise, you will need to define for the programmer in great detail how the process should operate. This definition should be understandable by all (operators, engineers, maintenance personnel, designers, etc.). The definition should be a living document that serves as your documentation for how the process operates. It also then becomes a reference tool for the operators (along with the SOPs) for maintenance and for others.

A clear, concise definition of the requirements is not made up of flowcharts and P&IDs. The specification must be written in plain language so everyone who has knowledge of how the process should operate can understand it and provide input. Many companies have also gotten caught up in talking about "structured language" to help define the specification. The words and structure that are used to write the specification are much less important than the fact that it is written and that it is clear.

The only way that this document can make sense is if the process is subdivided into functional units. The operation of these units can then be defined relatively independently of each other. Each unit can also be defined with a high level of detail.

Always remember that the person programming the control system is unlikely to be the person who maintains it. This makes the operational specification even more important. The operational specification should be the source code documentation. When done properly, the writing of the operational specification should take about 40% of the automation effort. With this done, the configuration and programming is a straightforward effort that will take only 30% of the time. Testing and start-up will then be much easier, because you know how the process is supposed to operate based on the specification, and should only take 20% of the time.

However, a typical project ends up taking twice as long as expected. In that case, the definition may take, optimistically, 10% of the time. The engineering and rework takes 60%. Start-up is a nightmare and takes 30%, at which time it is dropped because the project is out of budget. The automation then doesn't work as desired and is only partially used.

The automation definition should begin with how the equipment is laid out. The organization of the PFDs, the type of equipment used, and the modular breakdown of the process are all important parts of the automation definition. Early in the design cycle, the automation definition should simply define what the process is required to do and try to answer the following questions:

- What major processing steps must be accomplished?
- What modules or units will be necessary to perform these processing steps?
- What operations or units of work must these modules be capable of performing?

There are several success factors for the operational specification:

- The definition must be complete enough that a person knowing nothing about the process could program or troubleshoot the automation.
- The definition must not be redundant. Information must be defined, stored, and managed in one and only one place.
- The definition must be equipment-centric.
- The automation specification must focus on where the automation can add value.
- If the process is in a regulated environment, the definition must be a controlled document.

- The definition must be both understandable and accessible to the operators who run the process.
- More is not necessarily better. The goal is a clear, concise, complete, and unambiguous definition. Everyone needs to come away from this phase of the project with the same understanding of how the process will operate.

Process design and Automation design must be aligned. This means that things like the P&IDs must be designed with the same groupings as the automation. In other words, all elements of a unit should be on one P&ID whenever possible.

The entire design cycle should be done from an equipment-centric perspective (Figure 6.1). The automation is then part of each equipment entity. It is not part of the product. Each product recipe uses equipment entities (including the automation) to make a desired product. Everything about the operation of the process must be defined eventually. This definition has several parts. The first part is the definition of the physical process, how the process is organized and what its capabilities are.

The physical process is defined from the top down (Figure 6.2). That means that the requirements for the entire process are defined first. This is a very general description of the process: what equip-

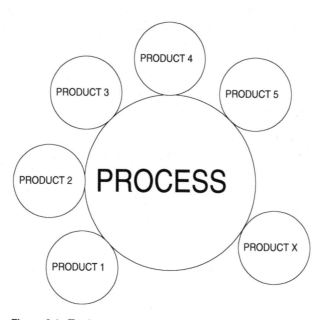

Figure 6.1. Equipment-centric automation.

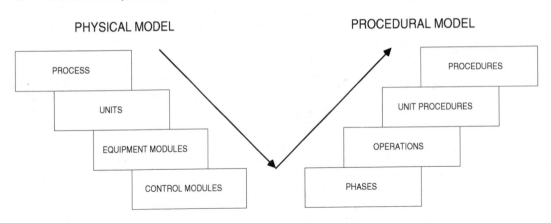

Figure 6.2. Modularization steps.

ment is needed, what major processing steps need to be performed, etc. This is written at the same time that the PFD is developed. Once the general process description is complete, the individual modules outlined in the description and in the PFD (usually units) are described in more detail. Then the same is done for each equipment module and control module. These are all defined in terms of the requirements defined at the higher level. The capability of each module is the summation of the capabilities of its submodules. In other words, the capability of a reactor unit is derived from the capabilities of all of the equipment modules (headers, transfer system, jacket temperature control, etc.) and control modules (agitator, pressure control, etc.) within it.

After the entire physical process is defined, the procedural requirements are defined. The procedural requirements use the information that was defined in the previous steps. The processing requirements that were defined earlier are now used to define the equipment phases: what work the unit or equipment module needs to perform.

Once the phases are defined, equipment operations, equipment unit procedures, and equipment procedures can be defined. These describe how to make a product using the process. As we discussed earlier, there are two types of procedural entities: equipment and recipe. Equipment procedures are usually specific to a given piece of equipment or module. We will discuss this definition in more detail in Chapter 9.

Equipment phases are the bases of all procedural control. For a given process though, equipment operations, equipment unit procedures, and equipment procedures may also be developed. Very rarely are equipment procedures used. They are typically used only in a process where one product will be produced and where a batch management system is not used. All procedural control is defined from the bottom up, starting with equipment phases.

There are four important documents that are created during Phases II and III of a project:

- Operational specification
- System specification
- Operational manual (SOP)
- Automation qualification

The operational specification defines, in a platform-independent way, how the process should run. It shows all basic, coordination, and procedural control requirements. It defines the alarm management scheme. We will spend much time throughout the rest of this book discussing the operational specification.

The system specification documents how the operational specification will be implemented in the control system. It starts out with the system requirements needed to achieve the desired automation results.

The operational manual is part of the SOP for the process. It documents how the operator interfaces with the automation and process to make product. The automation qualification documents how both the system and the operation will be tested. It documents the critical control parameters and the acceptance criteria for the automation. Whenever a change is made, it needs to be validated at some level, through the execution of another protocol or qualification. The qualification documents are controlled documents that must be stored and maintained, but changes to the automation require new qualification protocols to be written and executed.

IRENE: "This all sounds great, but where do we start? There are a lot of things to define."

ALLEN: "That's true. We are in Phase II of this project. During this phase of the project, we said that we need to create the operational specification and the system specification and we need to begin working on the operational qualification and the operational manual."

DONNA: "Well I have a product recipe that we could use to start the definition."

ALLEN: "We will definitely need some product recipes, but let's wait until a little later in the project. Right now, we need to define the equipment requirements in a product-independent manner."

DONNA: "How are we going to define the automation independent of products?"

ALLEN: "That's a very good question. Remember earlier when we talked about the S88 models? One of the models is the physical model. That's where we are going to start. The physical model is totally independent of product. All of S88 is based on some principles that I like to call 'modular manufacturing.' Let's discuss that next."

7

Apply Modular
Manufacturing Principles

ALLEN: "Now it is time to introduce a concept that is key to the entire Strategic Automation methodology. I have made reference to the term modular manufacturing in my descriptions of the process life cycle. In order to really understand how the concept of modular manufacturing works, we must first answer the question 'what is a module or modularity?'"

What Is a Module?

Webster's dictionary defines a module as "a detachable section, compartment, or unit with a specific purpose or function." A module is usually a member of a larger set of similar or like items. Each member of the set has a specific function that it performs differently from the other members. However, there is a very standardized method for all the members to interface with each other. One of the best examples of modularity is in children's toys. LEGO®[1] bricks are a perfect example of a modular system. Every LEGO® brick has a very standardized means of interfacing with other LEGO® bricks; they connect together by studs found on each piece, which fit perfectly into tubes found underneath each piece. These studs and tubes are always spaced the same distance between each other on every piece,

[1]Registered trademark of The LEGO Group, used here with permission.

so that when designing a new piece, the designer has the freedom to create any functionality needed for the piece as long as he designs the interface to other pieces according to the stud and tube standard. Allen said that he played with LEGO® bricks as a child some 20–25 years ago and plays with his children's now. This stud and tube system has served so well that new and very ingenious LEGO® element designs are still compatible with LEGO® bricks of yesteryear. Each new member piece of the whole LEGO® system of play has a unique function, allowing the designer and builder to create even more realistic and modern models; as long as they still use that same member-to-member stud and tube interface, they all are interconnectable. The idea behind this modular construction system is that you can use the same parts, each with unique functions but all with a common standardized interface, to build literally anything you can dream up (Figure 7.1).

It is very important to understand this simple example of modularity because it will be a central key to the success of the automation methodology. Each LEGO® brick acts as a module. It can be completely isolated from the other pieces and still perform its designed function, whether that function is just to be a square block with studs and tubes, or, a rocker arm assembly with stud and tube connection points. However, each piece can be connected to any other piece to form a completely different member subset module with completely different functionality than any of them alone. These pieces can then be connected to other pieces using this standard stud and tube interface to create even larger modules, or can be disassembled and reconnected differently to make a completely new subset. The key is that the basic modules are always there and never lose their independent identity, functionality, and interconnective capability as a module.

So far, we have defined a module as a unique member of a larger group of items with a common interface to each other that allows any

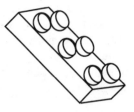

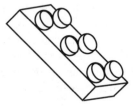

Figure 7.1 LEGO® bricks.

module to connect with any other module to jointly perform a greater combined function while still maintaining their individual unique definition and functionality.

ALLEN: "Can anyone give me another example of a module or modularity?"

DONNA: "The elements on the periodic chart are modules. They all have a very unique identity and the atomic structure of the electron rings is the standardized interface they use to join together to create larger modules. They retain their unique identity when joined together, in that when the larger module is subjected to the right conditions, the elements can be separated back into the individual smaller modules."

MARK: "Mechanical piping is modular just like LEGO® bricks. There are standard pipe sizes and flanges with standard bolt patterns that connect together. You design a pipe run knowing that by using the standard sizes and flanges you can interface to valves, pumps, and vessels using those same size standards."

ALLEN: "That's right, and you can also disassemble those pipes and valves and pumps and vessels into their original pieces, which retain their own unique original functionality."

IRENE: "This is like the new programming methods called object oriented programming. Isn't a module just another name for an object?"

ALLEN: "That's right, Irene. There is a lot of similarity between the concept of object oriented programming and modular manufacturing. In fact, when we talk about how we program the control system, we will see that employing the principles of object orientated programming (OOP) allows us to fully implement the modular manufacturing methodology. We will use some of the terms from OOP to describe modular manufacturing. For now, let's see how this idea of modularity applies directly to manufacturing and why it is important.

The primary reason for using a modular approach in manufacturing is to answer the age-old question of how to approach a large project. The answer is one task at a time! I could have just as easily said, 'in small, bite-size modules.' The idea is that anything is easier to understand when you can break it down into its smaller subparts. When setting out to build something, it is much easier to begin by separating it into its smaller subparts and design them so that when fully assembled they achieve the desired result. This end result is the sum total of each unique and independent subparts or module. "

PATTY: "Well, Mark has already pointed out that a chemical process is made up from modules of piping, valves, pumps, vessels, and other modules of equipment. So, how is this idea of modular manufacturing new or unique?"

ALLEN: "You're right, Patty. The ironic thing is that while chemical processes are designed this way in the first place, somewhere in the design and life cycle of the process, the benefits of this modular nature are lost. Each valve, pump, or vessel is designed to perform a unique processing function. These unique modules are then put together in various combinations, using some standard interface methods. However, there is more than one interface involved in the full implementation of the process. That is where the operation specification comes into play. It is the tool that forces us to look at all aspects of the process and the automation of that process in a truly modular perspective."

IRENE: "I see what you mean. When the process equipment itself is designed and assembled as a process, it is thought of in a modular format with each module having a unique function and standard interface, but, when it comes time to actually put the instrumentation and control around that finished design, it is thought of as one single large process rather than individual modules."

Why Modular Manufacturing?

ALLEN: "Exactly! The processing equipment itself is actually modular. There are three types of manufacturing processes: continuous, batch, and discrete. You probably are all familiar with these terms and conjure up a mental picture of a process example to accompany each method. Batch and discrete processing have traditionally employed the concept of modularity for automation and control better than continuous processes. This was primarily due to the need to process material in defined quantities known as batches or lots in a defined period of time and then pass it on to the next piece of equipment. More recently, most manufacturing processes have begun to use modularity to take advantage of another major benefit—flexibility. Today, the batch and discrete process industries may use the same equipment to make many different products, one after another, just by rearranging the various process equipment modules."

DONNA: "Yes, that's true. The days of 'mass production' are gone. Why should we design and implement our processes so they can only be used to manufacture one product or product variation? Why can't we get to the point where we can quickly modify our process to make better product or a whole new product with the same equipment. We do it in research labs and on pilot processes. It seems that we could get to the point where our process design doesn't prevent us from making different products that the process equipment was actually capable of producing."

PATTY: "This is all fine and good, but how do we go about designing these 'modular' plants?"

ALLEN: "For the most part, a good starting point is your PFD. Most major pieces of equipment are part of a unique module. A module is the equipment—all of the instruments, pumps, valves, alarms, and procedures to operate it ."

ROBERT: "When you say 'procedure,' do you mean SOPs?"

ALLEN: "Yes, but I don't mean just the SOPs. I'm also including all of the automation associated with the equipment. In S88.01 terms, that means all of the basic, coordination, and procedural control associated with the equipment."

IRENE: "So all of the phases for a unit are part of the unit module?"

How to Modularize

ALLEN: "Yes. Now back to drawing the module boundaries. Figure 7.2 shows the criteria we talked about that will help decide the boundaries between modules. The first thing to consider is the purpose of the module and its components. The purpose of the component must be aligned with the purpose of the module."

MARK: "What do you mean by that?"

PATTY: "Let me take a stab at that. In our process, the vacuum system has a very different purpose from the distillation column, although they are connected to the same line. They are probably separate modules that may be used together."

ALLEN: "Exactly! And you used a very important word in evaluating

- **Purpose**
- **Use**
- **Portability**
- **Flexibility**
- **Independence**
- **Expansion**
- **Isolation**
- **Physical Process Constraints**

Figure 7.2 Modularization criteria.

module boundaries—*system*. Anything that can be called a system is probably a module. Try to think of each module as a self-contained system."

MARK: "How can a distillation column be self-contained. It can't do anything without being fed material, and it needs to send the separated material somewhere."

PATTY: "But for the purpose of separating that material, it is self-contained. So long as it is being fed material, and has somewhere to send the separated material to, it has everything it needs to perform distillation."

LARRY: "OK, then how does use differ from purpose?"

ALLEN: "Use defines how the equipment interacts with its neighbors. Is it a shared utility like the vacuum system, or is it an integral part of the process like the reactor? There are three categories of use: stand-alone, shared use, or exclusive use. Shared use and exclusive use modules are resources that are acquired by other modules when they are needed. Shared use can be acquired by more than one owner module at a time, whereas exclusive use can only be used by one module at a time."

PATTY: "I would think that the vacuum system would also be a good example of portability. If you mean what I think you mean, the vacuum system automation strategy could be used almost as-is in another process."

ALLEN: "That's exactly what it means. Can the entire module be reused in another process?"

DONNA: "Flexibility is exactly what I want! We don't want to design the process or the automation to limit what we can make."

LARRY: "I understand why a process needs to be designed for expansion, but how does that affect the automation?"

PATTY: "Think about the vacuum system. If we were to add a new reactor to this process in the future, we would want the vacuum system to be able to accommodate this. That means that we need to draw the boundaries around the vacuum system such that few, if any, changes would need to be made."

ROBERT: "Well, I'll have to say that isolation sounds good to me. If we can design this process so that a problem in one reactor doesn't cascade and shut down the entire process, that would be great. Many of our safety problems are caused by the fact that rather than isolating the problem, we shut the entire process down. That sometimes causes more problems than the original event."

ALLEN: "The point about physical constraints is relatively obvious. Are there physical reasons for grouping or not grouping certain equipment?"

Object-Oriented Modularity

The success of a batch process design is determined by how flexible the recipes written for the given equipment can be and still maintain the stable rigidity of unchanging phase logic. In order to achieve recipe flexibility, it is necessary to be able to rearrange the order of activity performed by the pieces of equipment that execute each production procedure. In order to have phase logic rigidity, it is necessary to have the phases act on pieces or groups of equipment which will always have the same function, relative to the production cycle, no matter where they occur in the recipe. In this way, a maximum degree of flexibility is afforded by being able to rearrange the execution of these recipe phases, without having to change the phase logic or it's interface to the production equipment. The purpose of modularity in batch process design is to understand and organize the activity of the equipment, into equipment entities. The understanding of how and why to modularize the batch process is absolutely critical to a good flexible recipe design.

We view a module as a separate entity in and of itself; it does not care how the rest of the world acts or looks but simply executes itself in a consistent fashion every time. Modularity indicates modules designed with a singularly common object type interface to each other. As we said before, LEGO® building bricks are a good example of this. Each brick is a self-contained module with a consistently interchangeable interface mechanism. When building something with LEGOS®, it is possible to make additions, changes, or whole new designs by simply rearranging the brick modules. The functionality of each module remains the same, irregardless of how it is connected to the other modules. Maximum flexibility allows a simple change in the design by removing only one brick module and replacing it with a completely different brick module, utilizing the common object type interface. This is a perfect example of objective modularity and can be related to the same quality in a batch process design.

During batch process design we are dealing with modules and procedural elements that combine to make an equipment entity. These modules are identified as units, equipment modules, or control modules. Then there are procedural elements that contain the execution logic of how each module of equipment will function to create a process. These modules are identified as phases. The successful batch process design must appropriately define what the ideal boundaries are for modularizing the equipment, so that the corresponding execution logic remains unchanged when the recipe changes.

Following a top-down approach to optimum modular equipment de-

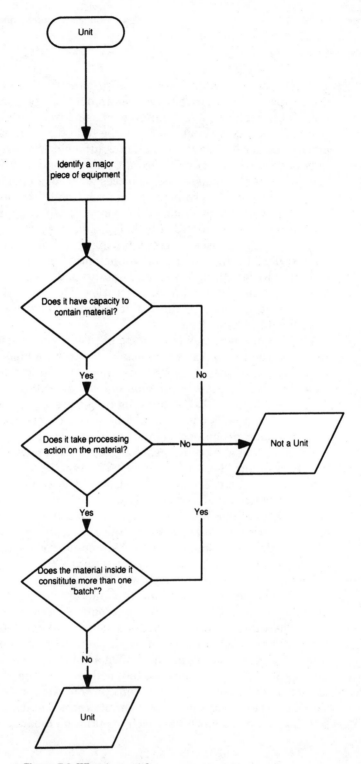

Figure 7.3 What is a unit?

sign, the following sections illustrate a flowchart method for evaluating a process.

Units

A unit is generally defined as a module of equipment that causes a change in the condition or state of the product. There are some rules of thumb that can be applied when deciding whether or not something is a unit. There will be exceptions to these rules and they need to be treated on a case-by-case basis. The rules of thumb are as follows (Figure 7.3):

1. Identify all the major pieces of equipment—things such as tanks, reactors, columns, and other vessels.

2. Does the vessel have the capacity to contain material in it for a period of time? A pipe has material in it, but not the same material for any period of time. If equipment does not contain material in it for a length of time, it is not a unit. An exception to this rule is a distillation column. It has limited (if any) capacity, but due to the fact that it changes the physical characteristics of the batch, it is a unit.

3. Does it have a processing action on the material? During the time the material is in the vessel, does it undergo a physical, chemical, or biological change? If the vessel just holds the material for a period of time, such as a hopper or storage tank, then it is not a unit.

4. Does the material inside the vessel constitute more than one "batch"? A unit can only process one "batch" of material at a time. Therefore, if the vessel does accumulate more than one "batch" of material at time, it is probably not a unit.

Equipment Modules

The most elusive level within the physical model is the equipment module (EM). Yet, equipment modules can have a high degree of impact on the flexibility and maintainability of the process (Figure 7.4).

1. Identify process actions (i.e., heat, agitate, or transfer out). These tasks identify what happens to the material. The tasks do not necessarily cause a change in the condition or state of the material. Generally, each one of these identifiable tasks will be executed by a single equipment phase.

2. Circle the equipment necessary to execute each task. These pieces of equipment are grouped together to create a preliminary equipment module. We now have the basis of a good modular design in that the independent phase tasks acting on the product can be executed with a uniquely independent group of equipment, independent of the rest of the equipment

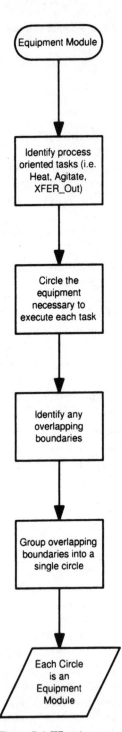

Figure 7.4 What is an equipment module?

or process. This provides modular objectivity, in that the scope of interface concern may be lessened when designing the activity of the EM.

3. Identify any overlapping boundaries. This step ensures that the design is objective as well as modular. To do this, we must look at any groups of equipment that share pieces of equipment. These groups must be consolidated into one EM with more than one equipment phase. This ensures that the module is completely objective, in that it can use its assigned equipment without regard to how any other part of the process is being executed. It is objective because it can be used to execute its phases completely independently of anything around it and it contains everything it needs to act upon the material, yet it has a consistent manner of interfacing with other units, equipment modules, and control modules.

Control Modules

The control module is the smallest module of equipment (equipment entity), and can be part of a unit, equipment module, or another control module. The definition of the Control Module provides yet another means modularizing the process to achieve the desired automation objectives (Figure 7.5).

1. Identify an output. The easiest control module to define is the field I/O output—equipment such as a control valve or block valve. Examples of control modules are a PID loop for temperature control and a block valve in a feed line.
2. Identify any measurements, alarms, and control actions associated with it. What will go on a loop diagram for this control loop?
3. Are there other outputs that should be associated with it? Is this a pump with a motor starter that has both a start and a stop output from the control system?

Example

ALLEN: We already started drawing some initial boundaries in our process. Now let us take a closer look at the process and continue drawing the boundaries."

PATTY: "Here is the block diagram that we talked about before (Figure 7.6). This is derived from the PFDs."

IRENE: "We already decided that the vacuum system and distillation column are both modules, and their boundaries look right to me."

MARK: "Based on the use criteria, the raw material 3 feed is a header. How would that be modularized?"

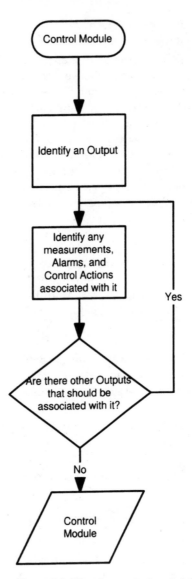

Figure 7.5 What is a control module?

ALLEN: "Do each of the reactors have their own valve and meter?"

PATTY: "Yes, they both need their own valve and meter. The header just supplies the material at a given pressure. Each place where the material is used needs to control the flow and amount added."

ALLEN: "Based on what you said, Patty, I would assume that both reactors can take material from the header at the same time?"

PATTY: "Yes."

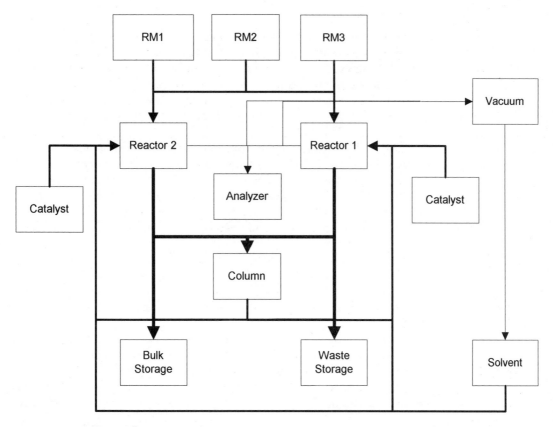

Figure 7.6 Sample process—starting point.

ALLEN: "Then, is there any reason for this to be grouped outside of the reactors, or can the raw material 3 addition be part of an equipment module within each reactor unit?"

MARK: "It doesn't look like there is any reason to have them outside of the reactor units."

ALLEN: "Since we are talking about the feeds, what about raw materials 1 and 2? How are they used?"

PATTY: "Raw material 2 is a tank. Both reactors can load this material at the same time. Raw material 1 is a bag dump station for loading dry materials. This can only be used by one reactor at a time."

IRENE: "It seems to me that both of these are distinct modules. Would they be units or equipment modules?"

ALLEN: "The difference is based on whether there will be a unit procedure written for each raw material module or if they will be acquired by the reactor unit and used as resources. We will discuss these differences later. For now, all we need to know is that these are each separate modules outside of the reactors."

LARRY: "The reactors will be a unit won't they?"

IRENE: "The reactors can't be in the same unit. They each have to be in their own unit, don't they Allen?"

ALLEN: "That's right, Irene. Since each reactor can be operating on a different batch at the same time, they need to be separate units. A single unit can only operate on one batch of material at a time."

PATTY: "Based on what we said about shared equipment, the analyzer needs to be a separate module also, right?"

ALLEN: "Yes. It is shared and used by the other modules when needed."

MARK: "Should we group the storage tank and waste tank into one module? They have the same purpose—storing material."

ALLEN: "Before we decide that, is the waste tank always going to store waste? Is there a possibility that at some point it might be used to store product?"

DONNA: "Based on what we are going to be doing with this process, who knows what will be stored in it. It very possibly might store product. Maybe we should just call it a second storage tank."

PATTY: "Once again, these two tanks can have different batches of material in them. Based on that, we need to make them separate modules."

IRENE: "What about the solvent storage? Is it a separate module, or is it part of the vacuum system?"

ALLEN: "Is the purpose the same for the two modules?"

PATTY: "No. The purpose of the vacuum system is to strip the solvent from a reactor. The purpose of the solvent storage tank is to store solvent and add it to a reactor."

IRENE: "Then these need to be separate modules, also."

PATTY: "It looks like we are done with the modularization."

ALLEN: "Not so fast. First of all, this is an iterative process. This is just a first pass. We also have to modularize the components of each of these modules. Is there anything missing? Are there any other parts of the process that aren't shown here? Some common components that may be missed are clean-in-place (CIP) or other utilities."

PATTY: "We *will* have a CIP system. It's not really part of the process, though. I didn't realize that we needed to define it here."

ALLEN: "It will be automated though, right?"

PATTY: "That's the plan."

ALLEN: "And it will interact with the other parts of the process.

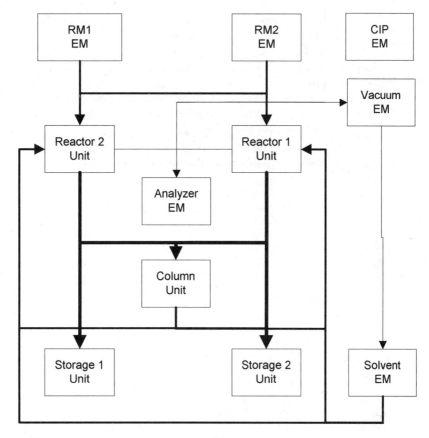

Figure 7.7 Sample process—block diagram.

Thus, it needs to be defined in the same way as the rest of the process."

PATTY: "Well, here is what we've come up with" (Figure 7.7).

Summary

LARRY: "This is all fine and good for a new process when the P&IDs aren't drawn yet. What if they are already complete and the process is mostly designed and you are just adding things to the process? How do you modularize then?"

ALLEN: "It is more difficult, but still possible. You still need to draw a block diagram for the entire process and apply the same modularization principles. You almost want to ignore the existing P&IDs and design it as if it were a new process. Once the process is modularized and the boundaries are drawn, then redraw the P&IDs based on those modules. For now, let us concentrate on the process we are working on."

IRENE: "We've modularized the process, what's next?"

ALLEN: "We have only begun to modularize the process. Now that we have a first pass on the modularization, the next step is to begin defining the physical process in more detail. This involves breaking down each of the modules that we defined into smaller modules."

8

Process Definition

DONNA: "I don't understand what else we need to know to specify the automation."

ALLEN: "We said earlier that our automation objectives are to provide flexibility, regulatory compliance, and a detailed batch record. We now need to define the process and all of its modules to achieve those objectives."

IRENE: "When you say 'all of its modules,' are you referring to equipment modules (EMs) and Control Modules (CMs)?"

ALLEN: "Yes. We need to define what EMs and CMs exist and what they are capable of doing."

LARRY: "Is this a mystery? The process should run automatically. What more do we need to do?"

ALLEN: "Traditional control definitions focused on single-variable regulatory control. The design used an extended bottom-up approach. The system was implemented based on the needed functionality. When additional features were required, they were added on an ad-hoc basis. Since the automation strategy that we're utilizing outlines a modular control structure, it should be defined at a high level. This definition makes it easier for the automation to change with evolving needs."

VICTOR: "Are you talking about system design?"

ALLEN: "The structure I am talking about will help you to design a better system, but I'm talking more about designing the automation. The system is the tool that we use to achieve that automation. The structure is based on the S88 models that we talked about earlier."

LARRY: "Isn't this going to delay the detailed design process?"

ALLEN: "Not necessarily. This may actually help the design process. We know that functional partitioning of the process makes sense. The automation needs the same level of structure and organization. The S88 models and terminology will help us establish the structure better. The biggest advantage, however, of this definition is that it becomes a tool to communicate with vendors and other related users of the system. Patty can start defining recipes well before you have decided whose system you will buy. Irene can start designing the phase logic. This structure can also be used for your systems, both new and retrofit."

PATTY: "Do we always have to use the complete models? We may end up with equipment from OEM's and/or inherit equipment that contains its own controls. Do we have to force-fit these models?"

ALLEN: "S88 models are generic models intended to cover most batch control situations. Your operational specification should define what your plant model will be. We are going to define how your process will execute from an automation perspective. In some cases, these pieces of OEM equipment can be called EMs and left unchanged. Even if they need to change, you may be able to minimize the impact by putting a wrapper or interface in the system that makes it look like an S88 module without changing the underlying code."

VICTOR: "I think our information systems people will benefit a great deal from implementation of this structure. They will be able to better define integration of business systems to the plant floor. Can you tell us how to go about defining the process?"

Define the Physical Model

The most critical step of the entire automation project is the modularization of the process. S88.01 goes so far as to say:

> All control related sections of the standard assume that the process cell in question (both physical equipment and related control activities) has been subdivided into well-defined equipment entities such as units, equipment modules, and control modules. Effective subdivision of the process cell into well-defined equipment entities is a complex activity, highly dependent on the individual requirements of the specific environment in which the batch process exists. Inconsistent or inappropriate equipment subdivisions can compromise the effectiveness of the modular approach to recipes suggested by this standard.
>
> Subdivision of the process cell requires a clear understanding of the purpose of the process cell's equipment. Such understanding allows the identification of equipment entities that must work together to serve an identifiable processing purpose.

Chapter 4 described the physical partitioning of the process into process cells, units, EMs and CMs. This partitioning should be performed in a top-down manner. That is, the process should first be identified. Once its boundaries are identified, the units within the process cell are identified. Once they are identified, the EMs within the process cell and all units are identified. Finally, the CMs within all levels are identified (Figure 8.1).

Process Cells

The process cell can include "trains." A train is a collection of units and associated EMs and CMs that can be used to make a complete batch. The batch to be executed may not use all the equipment defined in the process cell. The units in a process cell do not need to be physically connected. Some of the benefits of defining process cells include:

- Predefined paths. If there are parallel units at any point, the choices are limited by the definition for automatic or manual allocation.
- Choice limitation. If a particular combination of units are recommended for making a product, even though many units within the process cell meet the equipment criteria, a process cell will limit the choices.
- Convenient grouping. Physical location may make certain groupings convenient.
- Limit operational access.

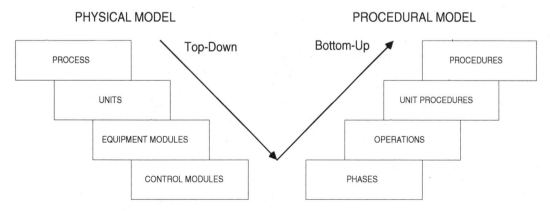

Figure 8.1 Modularization steps.

Units

A unit is made up of EMs and CMs. Some of these modules are permanently engineered to be a part of the unit, and some can be acquired as needed. Units should be able to operate relatively independently of each other. A unit can temporarily acquire the services of another EM or CM to carry out specific tasks. A unit will typically contain or hold the batch and cause some change to occur. By definition, a unit cannot make two products (execute two batches) at the same time.

Equipment Modules

- An equipment module (EM) is made up of other EMs and CMs.
- As with a unit, an EM can acquire the services of another EM or CM temporarily to carry out specific tasks.
- Since an EM can be a shared resource, it can be designed to operate on more than one batch at a time (e.g., vacuum system).
- An EM is capable of executing procedural control (phases).
- The primary difference between a unit and an EM is that an EM is typically called on by a unit to perform a specific task as opposed to acting on a batch independently.

Control Modules

- A CM is made up of other CMs and devices.
- Since a CM can be a shared resource, it can be designed to operate on more than one batch at a time (e.g., a header valve).
- CMs cannot execute procedural control. They are only called upon by procedural control to perform a specific action.

Define Tags

In defining units, decide on a tag naming convention. If units are grouped as a class, define a scheme that will consistently name devices. If dissimilarities are severe, consider moving the unit to its own class. This process will help in automatic tag generation by Control Systems. An example is level transmitters and an extension having the same number, thus differentiating them only by reactor number.

Define Resources

A resource is an entity (either EM or CM) that can provide services to a unit. Earlier we defined resources as exclusive or shared resources.

An example of an exclusive use resource is the analyzer. If another reactor wants to use the services of the analyzer, it has to wait for the other reactor to release it.

A shared resource can provide services to several units at the same time. An example of a shared use resource is a vacuum system. However, if the capability of the shared resource is limited, then arbitration of the resource becomes critical.

Arbitration is the process of resolving one or more requests for resource usage. The following are some ways that arbitration can be done:

- System based on schedule (batch priority)
- System based on availability
- Manually by the operator

After arbitration, a resource is allocated to a requesting entity. If there are no contentions or limits on capacity for a resource, then arbitration is not required and allocation is made directly. An allocated resource may need to be preempted when a higher priority batch requires the use of the resource. The process of preemption becomes complex, because a resource or unit allocated to a batch is removed and reallocated to another batch.

There are several approaches to arbitration. If the resources or units are true objects, then the arbitration can be made by the resource itself, acting on a message from another object. Arbitration can also be performed by an algorithm that manages all resources and units within a process cell. When the function of allocation and arbitration is done by the control system, it should be defined carefully. The units and resources should have states defined for allocation. These states should reflect operational reality.

Sometimes, some nonproduct recipes will be created such as clean-in-place (CIP) for the equipment. In a campaign mode, a unit may need to be sanitized after each run. At the end of the campaign, the status needs to be set to "dirty." If adequate states are defined for the equipment, then logic can be used to automatically set the status of the equipment. Then, during allocation a decision to allocate will be made based on this status and equipment state.

Define Modes and States

Procedural element modes define the manner in which a procedure moves from one procedural entity to another. Equipment entities may also have modes. The following are examples of typical mode definitions:

- Automatic. The procedure moves from one procedural element to another as long as the transition conditions are fulfilled.
- Manual. The procedure executes the current procedural element and stops. Further execution is determined by manual selection of the procedural element and is the responsibility of the operator.

A state describes the condition of an equipment entity or a procedure at any given time during its execution. Any procedural element (phase, operation, unit procedure, and procedure) can be in one of the defined states. Some typical states are running, holding, aborting, stopping, restarting, and complete. The equipment can also have predefined states, namely available/unavailable, open/close, remote/local, etc.

Modularization Criteria

Figure 8.2 outlines the criteria that can be used to determine the boundaries between modules.

Purpose

What is the intended purpose of the module? All of the elements of the module (inside the boundary) should have a similar purpose. A module should operate somewhat as a self-contained system. Make sure that everything within the boundary fits the system view. For example, the purpose of a vacuum system is to maintain a constant vacuum pressure on the inlet and pull vacuum on upstream customers. The

- **Purpose**
- **Use**
- **Portability**
- **Flexibility**
- **Independence**
- **Expansion**
- **Isolation**
- **Physical Process Constraints**

Figure 8.2 Modularization criteria.

purpose of a solvent recovery system is to condense solvent from a process stream and collect the solvent. The solvent recovery system may use the vacuum system to pull the solvent from a tank, but the purpose of the vacuum system is different from the purpose of the solvent recovery system. The vacuum system may be used independently of the solvent recovery system; therefore, they should be separate modules.

Use

Use refers to how the module interacts with other modules. Is it acquired by another module (unit or EM) or does it operate independently? If it is acquired by another module, can more than one module acquire it at the same time? Any module that can be acquired by more than one other module must be outside of the boundaries of both modules that can acquire it. For example, if a block valve CM can be acquired and used by two different reactor units, then the CM must be outside of the boundary of either unit.

Portability

Portability refers to the ability of a module to be duplicated or moved to another process or location. Might this module be duplicated in another process cell? Might this module be moved into another controller in the future? If so, then the boundary of the module needs to be appropriate so that it is self-contained and can be moved.

Flexibility

Is there a grouping of equipment that will increase or decrease the flexibility of the process? The modules need to be defined so as to not limit the flexibility of the overall process.

Independence

Independence refers to the ability of the module to operate on its own to the greatest extent possible. Sometimes that requires adding a recycle or reflux capability to the module. Independence is not always required, but it allows the process to operate much more smoothly.

Expansion

Expansion refers to the ability to add capacity to the process by adding new modules. If there is a portion of the process that can or will be a bottleneck, then it is usually a good idea to draw the module boundaries so that new capacity could be added by duplicating the module.

Isolation

Isolation refers to the ability to minimize the effect of process upsets by containing them within the module where the upset occurs. This goes hand in hand with independence.

Physical Process Constraints

A physical process constraint is any equipment limitation that may force modularization to occur in a certain manner. For example, if two tanks share a common pump, then the pump must be defined outside the boundary of either of the tank modules. The tank module then acquires the pump when it is needed.

Modularization for an Existing Process

A different approach should be taken when modularizing an existing process. When the process has already been designed, the P&IDs can be used as the starting point for the modularization. Whenever possible, the physical process and the automation should be designed at the same time.

Let us examine a sample process (Figure 8.3). This process consists of two premix tanks and two reactors. There are four raw materials that can be fed to some or all of the tanks. All of the tanks have some unique attributes. Based on the criteria that we have outlined, each raw material feed needs to be a stand-alone module outside of any of the tank units. Similarly, the feed inlet valves into each of the tanks need to be used by the feed modules to assure that the material is not fed to the wrong tank.

All of the control associated with each premixer and reactor is dedicated to the tank and therefore can be called part of the premixer or the reactor system. This includes the discharge pump, the valve from each premixer, and the inlet valve into each reactor. The cross transfer valve is used for transfers between premixer 1 and reactor 1, and it is also used for transfers between pre-mixer 2 and reactor 1 (Figure 8.4).

Example

PATTY: "You said earlier that we should draw the P&IDs at the same time that we define the operation of the process. Now you show us a

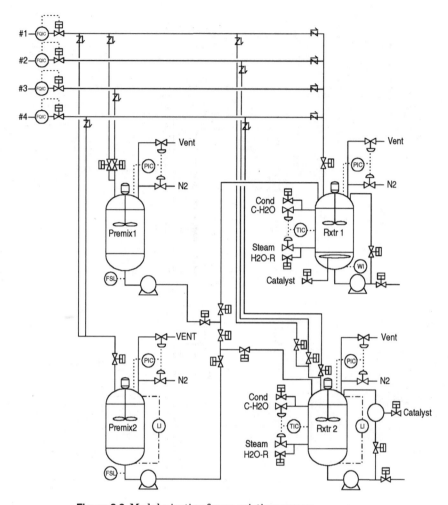

Figure 8.3 Modularization for an existing process.

modularization that circles elements of a P&ID. Which way is correct?"

ALLEN: "Alright Patty, you caught me. I highly recommend defining the operational specification at the same time that the physical process (P&IDs) is being defined. Unfortunately, in an existing process, this can't always be done. On this project, we have the luxury of doing both at the same time. On many projects though, they have to work with an existing process."

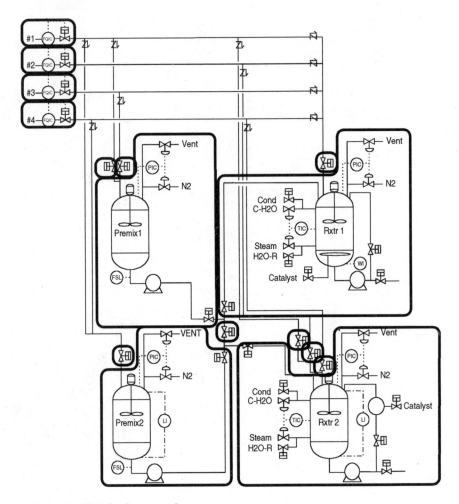

Figure 8.4 Modules for a sample process.

MARK: "So where do we start on this project?"

IRENE: "What do you mean; we've already started. We already drew a block diagram of the highest level modules" (Figure 8.5).

ALLEN: "Now let's take one of these modules, and modularize it further. How about one of the reactor units?"

PATTY: "Well, each reactor needs to heat; add material from drum, add raw materials 1, 2, and 3, add solvent, pressurize, agitate, pull vacuum, vent, recycle, and transfer out to another unit."

ALLEN: "Can any of these functions be grouped into a single EM?"

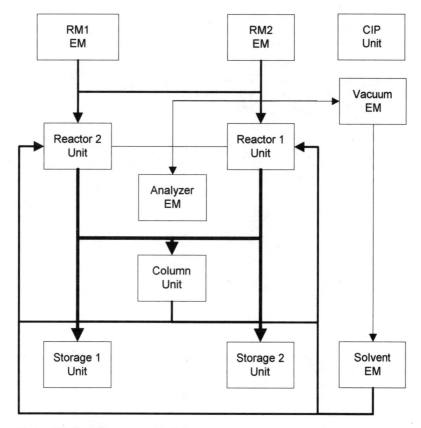

Figure 8.5 Example process block diagram

PATTY: "I guess that pressurizing, pulling vacuum, and venting will use some of the same equipment. Could they be a single EM?"

IRENE: "It certainly seems like it. The valves need to be aligned so that we don't pull vacuum at the same time that we are pressurizing. What about recycling and transferring out?"

MARK: "We are going to use the same pump for both recycling and transferring. What will we do about the shared equipment for raw material 1 and raw material 2?"

ALLEN: "In that case, the EM is outside of the boundary of the reactor unit."

IRENE: "So, if I understand this correctly, we will end up with four CMs (CIP feed, CIP return, drum addition, and temperature control) and four EMs (solvent, raw material 3, pressure, and transfer) inside the reactor unit. Is that correct?" (Figure 8.6).

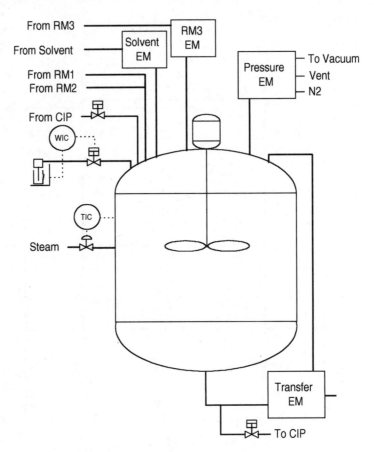

Figure 8.6 Sample process reactor.

ALLEN: "Now we need to go into more detail in each of the EMs and CMs. Then we need to do the same thing for each of the other units and EMs within the process."

PATTY: "Ok. So what do we do when we've finished defining the physical model?"

ALLEN: "That's when we start to define what each unit and EM is capable of doing through procedural control or phases."

Defining Equipment Procedures

IRENE: "Allen, you said that we need to start defining what each unit and EM is capable of doing? How do we go about doing that?"

ALLEN: "We need to define the equipment procedures that will run on the equipment. In most cases, the equipment procedures are equipment phases."

STEVE: "I'm still struggling with the difference between a recipe and a procedure. What is the difference?"

ALLEN: "That is a very good question. What do you think, Donna?"

DONNA: "Based on what you said before, a recipe contains a procedure. It also contains other information, like the formula, that will fully describe how to make the product. What I don't understand is the difference between an equipment procedure and a recipe procedure."

ALLEN: "OK. Let's discuss some of the differences and how to define equipment procedures."

Procedural Control

As discussed earlier, a procedure is the strategy for carrying out a process to make a product. What does that mean? In simpler terms, a procedure is the work that is performed by a process in order to make a product. Procedures include such elements as heat, cool, mix, transfer, separate, and react (Figure 9.1).

There are many alternatives to the mapping of procedural elements and equipment (Figure 9.2). For example, there may be units, equipment modules, and control modules within a process cell. An equipment phase may exist within a unit or within an equipment module. An equipment phase may act on an equipment module(s) or control module(s).

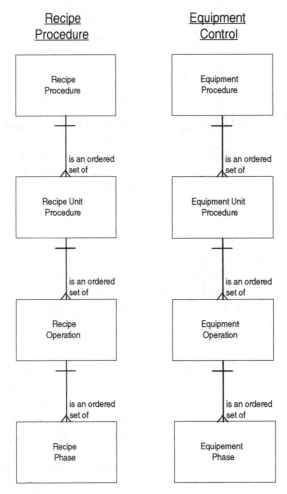

Figure 9.1 Procedural control hierarchy.

There are two types of procedural elements: recipe and equipment. Each of the levels (procedure, unit procedure, operation, and phase) can be either a recipe procedural element or an equipment procedural element. The difference lies in the independence or association of the procedural element to the equipment. A recipe procedural element is independent of the equipment on which it will execute. On the other hand, an equipment procedural element is specific to a piece of equipment.

If batch software is to be used to manage recipe execution, all of the procedural elements in the batch engine are recipe procedural ele-

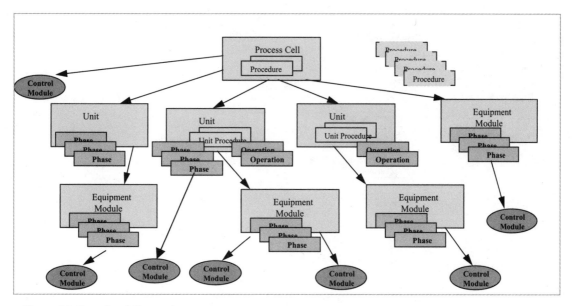

Figure 9.2 Procedural alternatives.

ments. All of the procedural elements in the control system (DCS, PLC, and PC) are equipment procedural elements. At some point, these recipe procedural elements and equipment procedural elements must be linked. S88 allows these elements to be joined across any level (i.e. from, recipe phase to equipment phase). Most batch software packages make the link between recipe procedural elements and equipment procedural elements at the phase level (Figure 9.3). In this case, the equipment phase is the only procedural element that is specific to the equipment. These equipment phases will typically run in the control system. The recipe phase, on the other hand will run in a batch engine on a supervisory system. Through this mapping of recipe phases to equipment phases, a heat recipe phase linked to two very different reactors will run its individual heat equipment phase. The end result on the product will be the same, however—the product will be heated to a given temperature for a given period of time.

On the other hand, in a process where a batch management system will not be used, there typically will be a different solution. This will be the case in either more-continuous processes or in processes that have few or only one product(s). Since there is no batch engine, there are no recipe procedural elements. That means that all procedural control must be done with equipment procedural elements (Figure 9.4).

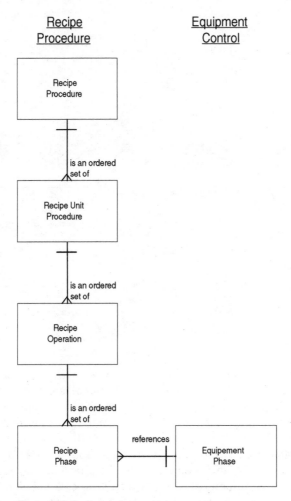

Figure 9.3 Procedural mapping.

Modes and States

Each equipment procedural element will have a number of distinct modes of operation. These modes need to be defined. Typically, the modes will be auto, semiauto, and manual. In auto mode, the equipment procedural element will be executed by either a higher level procedural element or a recipe procedural element. Manual mode allows an operator to acquire the phase and execute it manually, which is different than manually operating the process. In manual, the procedur-

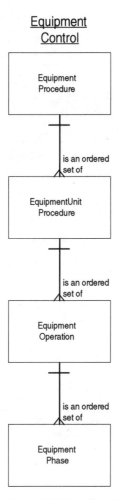

Figure 9.4 Procedural hierarchy for a continuous process.

al element will still run the process, but the operator chooses what to do, as opposed to manually starting and stopping pumps.

There are also a number of states that a procedural element can occupy. A typical state diagram for a phase is shown in Figure 9.5. See Appendix E for a complete description of the equipment phase interface.

A given equipment procedural element may have some or all of these states. It may also have other states not defined here. No matter what the states, they must be defined, and the methods of moving between these states must be defined. The "real" work of the phase is

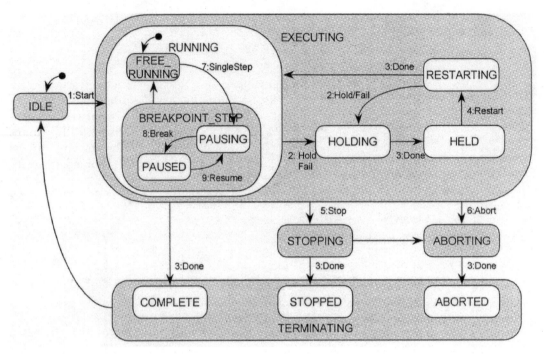

Figure 9.5 Equipment phase—state model.

performed in the running state, whereas the rest of the states are for exception handling.

Defining an Equipment Phase

When setting out to define an equipment phase, there are several things that need to be included: steps, transitions, parameters, alarm management, messages to the operator, communication to other entities, permissives, and interlocks.

A phase is a sequence of steps and transitions used to perform a process oriented task. Each step can perform one or more functions, such as opening valves, starting pumps, or resetting timers. Between each step is a transition. The transition defines what conditions must be true to move between steps. Typically, at the phase level, steps will not be performed in parallel; however, the steps do not necessarily need to be performed in order. Within a phase, there can be decision branches and looping structures when needed.

There are many different ways to define what work is performed during a phase. We recommend a combination of text, SFCs and tables. Figure 9.6 shows a sample table that can be used to build a phase. As you can see, this gives a description of the work to be performed by the phase, a graphical view of the work, and a tabular detailed breakdown of the work.

By preparing a tabular view of the phase, it is less likely that things can slip through the cracks. For each step, you will be looking at all of the CMs that can be manipulated. You will define what recipe parameters are used and how. You will show how the alarms will be managed. All alarms owned by the unit or EM are shown, and can be manipulated in any step. They can be enabled, disabled, priority

PHASE NAME

Description of the work to be performed by the Phase.

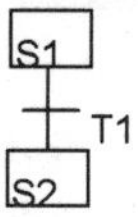

	STATE	RUNNING		STOPPING	RESTARTING	ABORTING	HOLDING
	STEP	S1	S2	S1	S1	S1	S1
REQUESTS							
Acquire Valve X		X					
INTERNAL RESOURSES							
Feed valve		Open		Close		Close	Close
ACQUIRED RESOURCES							
Valve X			Open	Close		Close	Close
ALARMS							
Alarm Y		Enable		Disable		Disable	Disable
TRANSITIONS							
Time > P1		T1					
Valves closed				Complete		Complete	Complete
MESSAGES							
PARAMETERS							
P1 – Time							
REPORTS							
R1 – Actual time							

PERMISSIVES AND INTERLOCKS

CONDITION	TYPE
Not full moon	PI

P: PERMISSIVE I: INTERLOCK

Figure 9.6 Phase definition template.

changed, etc. Every message to the operator is defined and associated with a specific step. The permissives and interlocks for the phase are also defined. The permissives are what allow the phase to operate. The interlocks are what will cause the phase to fail to a hold state.

For each of the active states, we need to define what the equipment phase will do:

- What communications will it send to a batch management system (requests)?
- What messages will it send to internal modules (CMs or EMs)?
- What resources will it acquire and what messages will it send to them?
- What will it do to manage the alarms?
- What will cause it to move between steps (transitions)?
- What communications will it send to the operator (messages)?
- What parameters does it expect to receive?
- What values will it send to be the batch history (reports)?
- What permissives and interlocks exist?

When we talk about "internal modules," we are referring to EMs and CMs that are inside the boundary of the unit or EM. Notice that we do not talk about output devices. An equipment phase does not talk directly to output devices. It talks to other modules. For example, rather than energizing an output to start a pump, an equipment phase will send a message to a pump CM that will in turn energize the output. In this way, the pump CM is an object, and owns all data and methods (i.e., start pump) associated with that pump. This allows different phases to acquire a CM (usually at different times through allocation and arbitration) and send messages to the pump CM without the underlying CM changing.

Transfers

A transfer is a unique type of procedural control. During a transfer, there is need for coordination across units. There are two ways in which a transfer can be handled.

1. Have a transfer out (XFER_OUT) on one end and a transfer in (XFER_IN) on the other end. Each unit controls its own resources (valves, pumps) to achieve the transfer. The units then communicate (usually through a batch management system to coordinate the transfer) (Figure 9.7).

2. Have a supply EM that has a phase to control the transfer. This can be

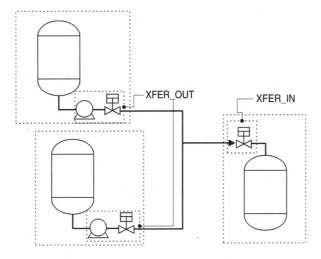

Figure 9.7 Phase coordination transfer pairs.

done with one phase for simple situations (feed system feeding two reactors). In more complicated processes, a phase is required for each combination of sending and receiving units (network of reactors and storage tanks) (Figure 9.8).

Define Phase Structure

A phase is the lowest level of procedural control. A phase performs process actions and requires parameters and procedural logic. The phase should be modular and follow a predetermined structure, so that it is easy to implement and document. At first, it may seem like a lot of work, but remember that this is the foundation of equipment control.

In defining the phase structure, the following considerations should be taken into account:

- State. Use predefined states consistently. For example, holding is always exception logic.
- Commands. Define and follow a set of commands to phases. For example, a Start command moves a phase from idle to running.
- Mode. Define what modes will exist for phases, and how they will respond when in each mode. For example, consider whether to have a single-step mode for troubleshooting phases.
- Exception handling. If the phase should do exception handling for errors, define the exception handling and recovery mechanism.

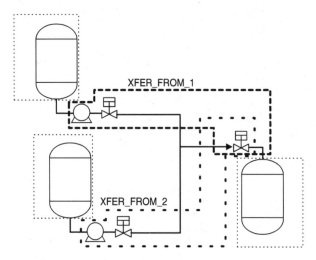

Figure 9.8 Transfer equipment modules.

- Messaging. A phase should provide messaging to inform the operator about the execution of the phase.
- Interface to basic control. The phase should not perform basic control, but interface with basic control—treat a basic control device as an object, and pass messages to it. In this way, EMs and CMs can manage their own mode and state. This will simplify changing and maintaining the code. If your control system does not treat devices as objects, consider building a wrapper around each device to make it more like an object.
- Process interlocks. Provide the status of process interlocks, which affect the phase, as messages.
- Programming method. The phase should be designed as a state machine. This will help in selectively enabling alarms and devices.
- Alarms. A phase should also treat alarms as objects. It shouldn't generate alarms directly. The phase could take corrective actions or perform exception handling based on alarm conditions.
- Parameters. Design the phase to be as flexible as possible by passing parameters.
- History. Design the phase to collect information for production history.

Example

IRENE: "Let's work through one of these phases. How about a simple one, the mix phase in one of the reactors."

ALLEN: "Sounds good. Patty, what does it need to do?"

PATTY: "Well, both tanks have a variable speed agitator. It just needs to turn on the agitator and mix for a period of time."

ALLEN: "Will it always be for a period of time, or could it mix for an indeterminate amount of time? How about the speed, shouldn't speed reference be a parameter?"

PATTY: "Typically, mix will be for a defined amount of time. I don't know whether the speed reference needs to change "

DONNA: "We do have products that need to be mixed until final assays are done in the lab, though."

ALLEN: "Then why don't we make the time to mix a parameter and allow a value of zero to mean 'mix indefinitely'?"

IRENE: "That sounds good. So would the definition look like this?" (Figure 9.9).

ALLEN: "That looks pretty good to me. How about a more complicated one now. Any ideas on a phase that has more parameters, needs to coordinate with another phase, and has more steps?"

MARK: "It seems like CIP is one of the most complicated parts of the process."

PATTY: "That's probably true, but why would we put the CIP system into a batch management system?"

ALLEN: "Isn't a CIP cycle a batch?"

MARK: "It certainly seems like it."

ALLEN: "By making the CIP system part of the standard automation, you can more easily manage the allocation and arbitration of equipment and schedule the cleaning of equipment as a normal batch."

IRENE: "I'm glad that we are going to talk about CIP. That's one part of the process that confuses me."

MARK: "It isn't really that bad. The CIP system cleans one tank circuit at a time. Once each cycle, it does the following:

1. High-speed wash. It washes the circuit with water and detergent for a set amount of time.
2. Rinse. It rinses the circuit with water for a set amount of time.
3. Dry. It dries the circuit with N_2 for a set amount of time."

IRENE: "Ok. That sounds simple enough, but does it reuse any of the water?"

MARK: "The rinse water is mixed with detergent and used as the wash water on the next cycle."

IRENE: "And what if there isn't enough or there is too much?"

MARK: "There typically won't be enough, and in that case more water is added to the CIP tank with detergent ratio controlled. If there is too much, the excess is sent to waste."

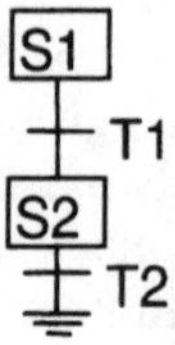

	S1 ┬ T1 S2 ┴ T2					
STATE	RUNNING		STOPPING	RESTARTING	ABORTING	HOLDING
STEP	S1	S2	S1	S1	S1	S1
REQUESTS						
INTERNAL MODULES						
Agitator	Start	Stop	Stop	-	Stop	Stop
EXTERNAL MODULES						
ALARMS						
TRANSITIONS						
(Time>P1) and (P1>0)	T1					
Agitator stopped		T2	Complete		Complete	Complete
MESSAGES						
PARAMETERS						
P1-Time to Mix						
P2-Mix Speed						
REPORTS						
R1-Actual time mixed						

PERMISSIVES AND INTERLOCKS

CONDITION	TYPE

P: PERMISSIVE; I: INTERLOCK

Figure 9.9 Mix phase. The mix phase turns on the agitator and runs until it receives a command to stop.

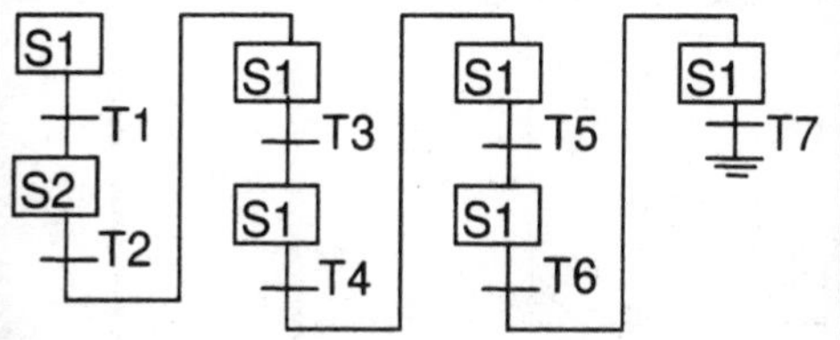

STATE	RUNNING							STOPPING	RESTARTING	ABORTING	HOLDING
STEP	S1	S2	S3	S4	S5	S6	S7	S1	S1	S1	S1
REQUESTS											
Ready to Clean unit	10										
Done cleaning unit							11				
INTERNAL MODULES											
RO water to process valve CM				Open	Close			Close	-	Close	Close
N2 valve CM					Open		Close	Close	-	Close	Close
CIP tank outlet valve CM		Open	Close					Close	-	Close	Close
CIP tank pump CM		Start	Stop					Stop	-	Stop	Stop
CIP tank process return valve CM				**		Close		Close	-	Close	Close
Waste valve CM		Open	Close	***		Close		Close	-	Close	Close
RO water to CIP tank valve CM		*	Close					Close	-	Close	Close
Detergent Control Module CM	Auto- SP							Close	-	Close	Close
Vent valve CM					Open		Close	Close	-	Close	Close
EXTERNAL MODULES											
ALARMS											
TRANSITIONS											
Unit ready	T1										
Time > P1		T2									
Valves closed			T3				T7	Complete		Complete	Complete
Time > P2				T4							
Time > (time to drain lines)					T5						
Time > Dry time						T6					
MESSAGES											

(continued on next page)

Figure 9.10 CIP phase.

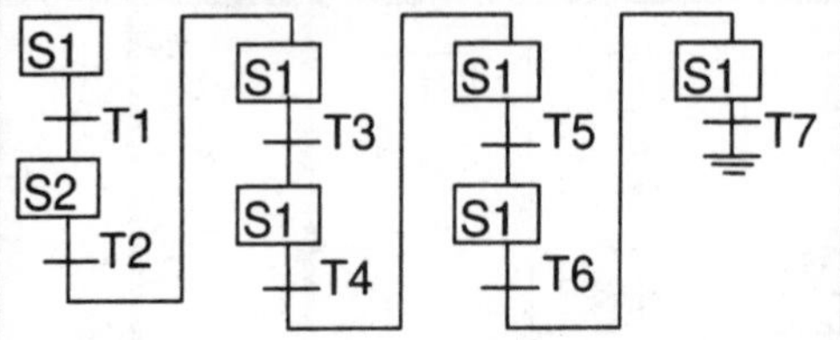

	RUNNING							STOPPING	RESTARTING	ABORTING	HOLDING
STATE											
STEP	S1	S2	S3	S4	S5	S6	S7	S1	S1	S1	S1
P1 Time to wash											
P2 Time to Rinse											
P3 Time to Dry											
REPORTS											
R1 Actual Wash time											
R2 Actual Rinse time											
R3 Actual Dry time											

*Open if level in CIP tank goes below 10%

**Open until the level in CIP tank goes above 80%

***Closed until the level in CIP tank goes above 80%

PERMISSIVES AND INTERLOCKS

CONDITION	TYPE

P: PERMISSIVE; I: INTERLOCK

Figure 9.10 *(continued)*

OWEN: "Does the dry time include time to drain the lines?"

MARK: "I guess we should delay before starting the dry timer."

IRENE: "Allen, how do we coordinate this with the equipment being cleaned?"

ALLEN: "In most cases, a clean phase is used on the equipment being cleaned, and we allow the batch management system to coordinate the phases."

IRENE: "So would the CIP phase look something like this" (Figure 9.10).

MARK: "That seems to do it."

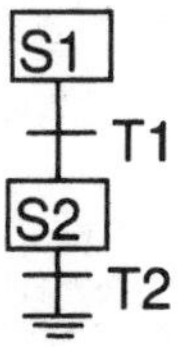

STATE	RUNNING		STOPPING	RESTARTING	ABORTING	HOLDING
STEP	S1	S2	S1	S1	S1	S1
REQUESTS						
Ready to clean	10					
Wait for done cleaning message		11				
INTERNAL MODULES						
CIP inlet valve	Open	Close	Close		Close	Close
CIP return valve	Open	Close	Close		Close	Close
EXTERNAL MODULES						
ALARMS						
TRANSITIONS						
CIP system ready	T1					
CIP system done		T2				
MESSAGES						
PARAMETERS						
P1 Temperature setpoint						
P2 Time to hold temperature						
REPORTS						

PERMISSIVES AND INTERLOCKS

CONDITION	TYPE

P: PERMISSIVE; I: INTERLOCK

Figure 9.11 Clean phase in a reactor unit.

ALLEN: "That does it on the CIP end. What about the other end?"

IRENE: "So if we are going to clean the reactor, we would have a corresponding clean phase in the reactor unit?"

ALLEN: "That's right."

IRENE: "Would it look like this?" (Figure 9.11).

ALLEN: "That's what I would do."

Summary

LARRY: "So you're telling us that we need to define all of that information for every phase?"

IRENE: "How else do you expect me to know how to program the automation?"

LARRY: "I don't know, but this seems like a lot of time and effort."

IRENE: "Believe me, it's less time than it would take for me reprogram the system several times."

STEVE: "And I'll know what the system is supposed to do during commissioning."

OWEN: "And it should work."

LARRY: "We've gotten all of the way to this point, and we still don't even know what system we are going to use."

ALLEN: "How about if we discuss the system next?"

10

Systems Strategy and Architecture

VICTOR: "We have defined the process and how it will operate. Now we're starting the design work on the project. I know that we all remember that we committed to finish this project in 28 weeks."

LARRY: "As you've seen, we have a discipline-based engineering team. The project team consists of a project manager, and several project engineers. Each project engineer is responsible for his or her area of specialty. The project team also includes the product team representatives. The product team is made up of a marketing specialist, a QA professional, a development scientist, and a business manager."

VICTOR: "How come operations is not represented in the project team? They should be in the future."

IRENE: "Where does the automation component of the project fit in?"

LARRY: "You're the project engineer for process automation. We've already completed the process definition. I assume that the estimates for the automation are valid?"

IRENE: "You mean the estimates we made a year ago?"

LARRY: "Yes, our capital for the project was allocated based on that estimate."

VICTOR: "Do we even know what systems we are going to use? We need the control system to be tied into our business systems."

IRENE: "I don't even know what level of integration is needed."

VICTOR: "We need to define what we need in a control system."

ALLEN: "Earlier we went through the exercise of translating the business goals to automation objectives. We need to establish a strategy to accomplish these objectives. Do you have a systems strategy?"

LARRY: "Why do we need a systems strategy to implement a single project. My job is to ensure that we get the best system for our money. It's always worked before."

VICTOR: "The information technology (IT) team established strategies for our business systems. We know that IT plays a key role in our business. Are you are saying that automation can play a strategic role in accomplishing the business objectives?"

ALLEN: "I consider manufacturing a key component of business. Automation is an enabler that helps run your processes. The system can generate meaningful information about your manufacturing operation. You need to have a consistent systems strategy. More so, because, rapid technological changes and an ad-hoc project approach has created islands of automation."

VICTOR: "OK, so I see the benefits of systems strategy. It seems like we need a consistent approach to automation systems."

ALLEN: "There are several benefits to having a systems strategy that works."

- Builds a consistent approach to solutions
- Identifies technology requirements
- Provides inputs for automation plans

VICTOR: "We need to ensure that we have a systems strategy in place and the infrastructure to carry out the strategy. Can you help us define systems strategy for automation?"

LARRY: "Do I have to pay for this effort out of this project?"

ALLEN: "If you have a strategy in place, which you follow consistently, it can help you save project costs in the future."

LARRY: "I don't see that happening."

ALLEN: "Did you know that about 20% of the definition time is spent in selecting systems and getting a well-trained team to start implementation for a project? If you have a system that has already been selected and you have the right expertise, you can save on selection costs and time for each project.

Now we should discuss systems strategy. I would recommend that this be carried out by a team. The systems strategy should be applied across an enterprise. I will outline the process of developing a strategy, but not the strategy itself. The systems strategy should be based on business goals. Can you list your business goals?"

VICTOR: "Our business goals for manufacturing are to make products of the highest quality at the lowest possible cost in the shortest possible time. We also have to make the products in a safe manner. Our company has a long-standing history of safety."

ALLEN: "Automation can help in meeting some of these goals. Technology advances have provided more user-friendly systems with more power and distributed data-handling capability. The development tools that are available today will help you with different design approaches."

When translating business goals to systems strategy, start with an analysis. The analysis should include where you are at present and a list of the approaches and challenges that need to be met. The result of this analysis will lead to requirements. The requirements need to be addressed as part of a plan. For example, the business goal of making products at the lowest possible cost could be accomplished by increases in efficiency or by reducing costs. This objective can be met through the timely capture of production data and the availability of tools for production personnel to analyze production data.

The contents of the strategy should be described in a simple and concise manner. Most system strategies have an implementation plan. The following issues must be addressed in a systems strategy:

- How the system will fit into the overall enterprise.
- What system(s) will be used.
- What approach will be used over the life cycle of the system (for development support, training, and maintenance).
- When and how will new technologies be evaluated?
- What level of consistency is desirable among systems?

After the systems strategy is developed, publish the document. Everyone within the enterprise who will be involved in decision making for systems should be knowledgeable about the strategy. The strategy document will have to be updated with the business strategy, and when changes occur, the systems strategy will have to be validated against these changes.

Automation Plan

An automation plan is developed to implement a systems strategy. The plan should state where you are at present and where you need to be. The plan should also detail how to get there. The plan needs to have a champion who can follow through. The elements of the plan may be implemented during projects, but the purpose of the plan is to provide a consistent framework for all process automation systems, new and old. Some of the key aspects of the automation plan are:

- Architecture approach
- Selection of vendor
- Set up integration lab
- Definition of standards
- Selection of integrator
- Toolkits

The Architecture

The automation plan should outline the architecture of the automation: hardware, operating system, communications, database, and application software. There are two broad approaches available for architecture. One is provided by vendors (single-window), and the other developed by users (make-your-own). Vendors today provide complete solutions for process automation, including integration into business systems. In this approach, all components are provided by the vendor. The other approach involves selection of the architecture by the user. In this approach, the components of the architecture are chosen at the enterprise level to ensure consistency and supportability. The technology choices have been evaluated and standards set. Any application that can adhere to these standards can be used. This approach will ensure better integration with other systems, however, both approaches are valid. The user architecture approach allows selection of technologies that closely match the business needs of the enterprise, whereas the technology choices have already made in the vendor architecture approach. The framework is developed and maintained by the vendor. The vendor may publish application program interfaces (APIs) for integration to other systems.

Select Vendor

Vendor selection is the activity of finding a vendor that meets your functional and technical requirements for a long-term relationship. Traditionally, vendor selection has been done for each specific project, with project specifications followed by a vendor's bid. This process should, however, consider the life cycle of the system. Considerable time and resources should be devoted to the vendor selection process. The best approach is to use a self-evaluation of the vendors, based on your established requirements. The format for the self-evaluation must be clearly defined with the responses being yes or no, standard or nonstandard. This will help to evaluate the responses. The selection process itself should be a team-based effort. A team brings unique perspectives together and balances out subjective biases.

The Requirements

The requirements should detail the technical, functional, integration, support, and maintenance requirements. To determine vendor viability, questions about vendor organization, development methods, and industry experience must also be included.

Technical Requirements

The technical requirements should list the preferred technical platform, operating system, standard database, network systems, and unique requirements (such as touch screens or monitors in hazardous areas). The specific operating conditions that the system needs to support should also be described. The requirements for a plant operating 24 hours a day, seven days a week will be different from a plant operating 12 hours a day. Redundancy requirements will be different for each operation. Specific application requirements to match the existing system should also be listed (i.e., reporting, historical data collection). The technical requirements must include some level of performance requirements. The performance requirements should be structured in such a way to bring out the limitations of the system. The performance requirements may include update times and throughput needs.

Functional Requirements

The functional requirements should first outline the functional roles of people who will interact with the system. The functional requirements should be based on the functional activities that are required to support making a product as seen from a systems perspective. For example, the functional requirements may be derived from the functional model of S88. Even diagnostic requirements must be listed here.

Life-Cycle Requirements

The life-cycle requirements should include cost, licensing, support, training, and maintenance-related requirements. The pricing structure for software varies more than hardware pricing. Obtain the complete pricing structure with breakdown on hardware and software prices. A fully working configuration may require other components, such as a relational database and hardware, which may not be supplied by the vendor. Ensure that pricing is included for evaluation.

Baseline configurations must be established to get a good comparison. The baseline configurations should be determined from previous experience. If a typical process plant has less than 20 units and 1000 devices, do not set a baseline configuration of 100 units and 10,000 tags. The application development costs should also be factored in here. Some systems may provide better configuration and application development environment than others.

Support requirements should include telephone support, on-site support, revisions, upgrades, and documentation outside of warranties. This may include system administration training as well. The mechanism for problem resolution and response method from the vendor should be requested.

Here are some life-cycle issues that should be considered when selecting a vendor:

- Pricing
- Annual support costs
- Upgrades, releases (costs, management)
- Documentation
- Installation assistance
- Support options
- Training (standard options)
- Remote support (phone support procedures, dial-in)
- Account management (marketing, technical contacts)

Ask the Right Questions

The selection process involves finding the right vendor. References play an important role in the decision process. Vendors typically offer their best customers as their references. When checking references make sure that they are of the same size and from the same industry. Also, ask questions about the capability of service organizations. Make sure that the vendor serves your industry. For example, a software vendor who develops applications in a particular industry may not have knowledge of a processing industry that is highly regulated. If the vendor will provide applications engineering, find out about project implementations. The size of projects must be comparable to what you might implement. Talk to application developers from the vendor organization.

Evaluation and Scoring

Evaluation of the vendor is best accomplished by first determining objective criteria for scoring. The scoring process should be organized in a structured format in order to help multi-criteria decision making. The format will break down the problem of selecting the vendor into its hierarchical components. Each component can then be broken down into increasing levels of detail. The detailed attributes should be organized into logical levels. This grouping of attributes helps in a rational evaluation and relative comparison across each attribute. After grouping the attributes, weights should be assigned to each attribute. The weights of the attributes within a level will equal one, making the components' relative weights equal to one. Then you can figure out how much an attribute contributes to the overall decision. For example, from Figure 10.1 you can infer that support contributes (0.2·0.25) or 5% of the overall decision. The weights may be altered for "what if" scenarios. During the scoring process, select and use a scale that will work for you. It can be a one-to-ten scale or one-to-three scale. Score all vendors for each attribute. Add up the scores and average them within each level. Multiply this average by the level weight to get a weighted score. Add the weighted averages and

```
Best Vendor (1.0)
    Vendor viability (0.25) [Level 1]        ·
        Vendor maturity (0.2) [Level 2]
            Number of product releases (0.2) [Level 3]
            Product history (0.2)
            Development methods (0.6)
        Vendor vision (0.15)
        Finances (0.45
        Customer base (0.2)
    Functional requirements (0.25)
        Recipe management (0.2)
        Historian (0.2)
        Batch management (0.3)
        Process control (0.3)
        User interfaces (0.1)
    Technical requirements (0.25)
        Hardware (0.25)
        Software (0.25)
        Interfaces (0.25)
        Development tools (0.25)
    Life cycle requirements (0.25)
        Cost (0.4)
        Documentation (0.2)
        Training (0.2)
Support (0.2)
```

Figure 10.1 Vendor evaluation.

then multiply by the weight for the next level. See the figure for relative grouping and weights; it shows the relative grouping of attributes in levels.

Demonstrations

After the scoring process, select the top two vendors and invite them to give technical presentation and demonstration. Their demonstration should be based on the specifications, including a process description and criteria for evaluation. The demonstration should use real-life scenarios that can be a part of the demonstration specification. Some of them should be unscripted, not published as a part of the demonstration specification, but helpful for understanding the usability and support requirements. For example, an unscripted scenario could include changing the name of a unit or adding a new attribute to a piece of equipment.

The Relationship

After this selection process, the vendor choice is finalized. A formal relationship, such as an alliance, agreement, or partnership with the vendor should be established. An agreement focuses on reducing the costs for both the user and the vendor. The user could have one or more agreements with vendors.

An alliance is an exclusive relationship existing over a long period of time, typically three years, and is based on solving a specific problem. The vendor could be the technology or resource provider. The solution is developed jointly.

In a partnership, the relationship again is exclusive, but information is shared and resources dedicated. A partnership provides a different perspective to the relationship. The relationship is established for mutual benefit and is long-term. The vendor could share future product plans or seek input for improving existing projects. The customer could share the automation plans. This could also lead to better communication between the partners through consistent channels.

Each relationship involves time and effort. The factors and outcomes involving the selection of the relationship should be well thought out. If a user is interested only in getting guaranteed short time deliveries with the least amount of paperwork, then a partnership may not be the most appropriate relationship. Trying to achieve a manufacturing objective not possible with conventional approaches may require a partnership. Institute benchmarks or mechanisms to measure the health of a relationship. Always remember that communication is the key to any successful relationship.

Integration Lab

The next step is to devise an integration lab. This is essential if a user-driven architecture is selected. The integration lab helps to prove concepts and test configurations. The integration lab can also be used for staging new systems or testing systems before installation. It can be also be used as a test bed for new technologies. The integration lab can use spare components, since it should not be a part of the production system. The system could also be used to provide a support framework.

Define Standards

Standards are important because they provide consistency in every phase of the life cycle of systems. There are different types of standards: technology, design, and methodology standards. Technology standards can help in assuring compatibility of disparate products. Design standards are desirable for things such as user interfaces, alarm management, and interlocks. These standards can provide consistency across an enterprise for automation. Methodology standards are specific to design, implementation, and testing of solutions. This group of standards may 1) outline a methodology 2) list principles of design, or 3) just offer guidelines.

The Standards Process

Standards should be developed through consensus whenever possible. This helps to get a better buy-in and to assure that all needs and perspectives are considered. You should clearly understand the objective of the effort before developing a standard.

Different types of standards shouldn't be mixed nor should guidelines be standardized. Standards should not be static; they should evolve and grow with changing needs. Make sure that current standards take future trends into consideration. Standards should be periodically reviewed to ensure that they remain current.

A project-based approach should be used when developing standards, listing the deliverables and work products. The standard should be published only after it has been reviewed and approved. However, the standardization process isn't complete until everyone is informed and educated on the standard. It is critical to get everyone to use it consistently. If it is a conformance standard, you must make sure that a compliance mechanism is in place. Most standards fail because there is no follow-through. A good way to get people to follow a standard is to showcase projects that have followed them successfully.

Select Integrator

A systems integrator helps in the design, implementation, and testing of projects and is required if in-house resources do not have the expertise or the time to complete the projects. Each company needs to uniquely define the role of the systems integrator. He or she can be brought in as an expert during earlier phases of the project or used as a resource during project implementation. Sometimes, the systems integrator takes full responsibility for the solution (the turn-key scenario). The selection process for the systems integrator can follow the same method for vendor selection outlined above. There are several factors that should be considered in the selection process:

- Project methodology
- Industry knowledge
- Technical expertise
- Flexibility
- Capability to deliver
- Accessibility
- Support

Project Methodology

A systems integrator should have methods and models for defining, implementing, and supporting solutions.

Industry Knowledge

The systems integrator must have a good knowledge of the industry, its best practices, and its regulations. Consider employing a systems integrator who specializes in your industry.

Technical Expertise

A systems integrator should be able to demonstrate expertise to analyze, design, implement, manage, and support a range of technologies.

Flexibility

A systems integrator should be flexible enough to adapt to changing needs. He or she must have engineering, integration, and project management expertise and should be able to work in multiplayer, multiorganization teams.

Accessibility

The systems integrator should be located reasonably nearby or be committed to having an adequate number of qualified resources on-site. Even with good electronic communications, information flow can be hindered if the systems integrator is located several hundred miles away.

Capability to Deliver

The systems integrator should have a record of delivering projects on a comparable scale in your industry.

Support

The systems integrator's capability to provide comprehensive and flexible programs for support should be critically evaluated even if you don't plan to use support services right away.

Define Toolkits

A well-organized workshop will have tools arranged so that a knowledgeable person can get the job done in an effective manner. With software, you need tools that apply to different functions, such as design, testing, and software administration. In software engineering, this full suite of functionality is sometimes referred to as a CASE tool. Toolkits are usually built to fulfill a need. The most popular tools are for design and analysis.

Application Toolkits

Application toolkits help in building automation software applications. The toolkit could be a part of the software that is provided by the vendor or created by the user. The nature of toolkits varies widely. Toolkits may include documentation templates, configuration files, or libraries for design implementation. Libraries can be built for display objects, phases, other recipe and equipment procedural elements, and control objects. Toolkits can take the guesswork out of implementation. Toolkits are usually produced for a standard configuration. Toolkits can be excellent vehicles for the delivery of standards. This is true for user interfaces, procedures for system administration, and database schemata. Toolkits can also play a key role in translating knowledge for others to use.

The toolkit approach is recommended for users who have multiple projects occurring at the same time. This approach will also help users with a lot of similar units. Toolkits can be used as a communication tool for both systems integrators and users. Toolkits help save valuable engineering time during the definition, design, and implementation phases of the project. Toolkits will become more important in the future as diverse technologies are used to support a manufacturing operation. Toolkits must be maintained and expanded as new requirements and features are added to new or existing installations.

Hardware

PLCs were initially used in discrete operations as a replacement for electromechanical relays. They have since evolved, gaining greater functionality, by adding networking and process control capabilities. Advances in user interface software have helped the PLCs as well. Supervisory control and data acquisition (SCADA) systems help in integration of plant floor controllers.

DCS systems were first introduced in the mid-70s to replace control panels. The systems allowed centralization of tasks but spread the risk of failure by distribution of functions. The systems could be configured, instead of being programmed to implement a control function. Today, DCS systems have evolved, taking on sequential and discrete control capabilities. The functionality is now available in both PLCs and DCSs. DCS systems are also adding traditional information management functions, beyond the normal control capabilities.

DCS systems are following an evolutionary path to keep up with changes in technology and demands of users. DCS vendors have started leveraging existing technologies or de facto standards to provide open systems. They provide some degree of freedom in selecting the best-in-class application software, databases, and networks. Each vendor uses a different approach. The result will be ease of integration with other applications and systems. For example, a DCS vendor may offer a choice of network protocols or databases.

There are also fundamental changes in the way distributed systems are built using object concepts. Systems follow established object frameworks, such as common object request broker architecture (CORBA). This concept helps in creating heterogeneous networked systems as long as the systems can exchange information using the same object framework.

When selecting the hardware platform for a specific application, several factors need to be considered. The factors may have different weights for each user. A large production-scale facility that processes hazardous substances may have higher requirements for reliability than flexibility. There are also situations where one system cannot do it all. User preferences play a key role in determining the choice of systems. Using a selection process outlined above will eliminate bias and help in a rational selection. The approach should consider a sample configuration to get baseline costs. Under functional requirements, consider attributes such as response time and alarm management.

Software

Software can be broadly classified into operating systems, middleware, and applications. An application is custom-written or configurable software that helps to perform automation functions. These functions are performed through operating systems and middleware.

Middleware is the software that is neither a part of the operating system nor a part of the application, but is required at run-time. This includes database management, communications, and interoperability middleware such as remote procedure call (RPC) and object request brokers (ORBs). Middleware plays a key role in cooperative processing (client/server computing). Middleware acts as the glue that binds application programs to components in an environment.

Middleware makes it possible to distribute application and data across multiple servers. With distributed object computing, software can query and acquire interface specs for application components at run-time. The capability for late-binding means links are created at run-time. This capability reduces the need for system management functions, allowing objects to connect to each other regardless of where they are located.

Microsoft Object Linking and Embedding (OLE) is a good example of how desktop programs are linked to one another. A spreadsheet application can be embedded as an object into a word-processing document. It uses Dynamic Data Exchange (DDE) as the protocol for messaging. ORBs introduce a new way of making applications interoperable. Distributed OLE will provide this functionality over the network.

Connectivity

The need to get production information will necessitate connection of automation systems to business systems. Most approaches have required some form of custom integration using a separate layer of software. The old hierarchical models of computer systems for manufacturing will give way to flatter models of just business systems and process automation systems. Connectivity of these systems will be a major issue since each has a different focus. Business systems are focused on return on investment (ROI) and profitability, whereas manufacturing systems are focused on control and optimization.

How Do You Protect Yourself?

In light of all the technology changes and rapidly changing business practices, our systems will be under increasing pressure to cope with change. Now more than ever, software plays a large role in manufacturing operations. Here are a few simple rules that can help protect your current and future investments in automation systems.

- Select an architecture that lends itself to growth
- Use a consistent approach on all automation projects
- Select best-of-class software applications
- Stay on top of technology

Select an Architecture that Will Lend Itself to Growth

The system should be able to grow with your changing needs. A system that is modular and scaleable will be easy to expand. The addition of a process unit should not mean that you must reconfigure your software.

Use a Consistent Approach to All Automation Projects

With a consistent approach to projects, the methods that you use are repeated. This ensures that project quality is maintained.

Select Best-of-Class Software Applications

If applications need to be purchased and configured, such as historians or batch management systems, select a vendor who has the best-of-class software. This will increase your chances of finding support. Some applications may require additional integration in order to work within your infrastructure.

Stay on Top of Technology

Changes are rapid in the field of computing. For instance, network computing has changed automation forever. By staying in tune with technology, you can make a rational assessment of when to invest.

Overall Trends

Process automation, like other manufacturing enablers, is at the cusp of a major revolution. Technology changes have helped processes to operate better and have redefined how businesses function. Some other key trends that will affect automation systems are identified below.

Networks

Networks have been integrating control systems for the past 20 years. Proprietary networks have been used in most of the DCS systems. This meant that things such as gateways and protocol converters were the only options for connecting foreign devices. Increasing user demand for interoperable systems have forced vendors to develop and/or adapt standard protocols. The microprocessor revolution has also helped allow the creation of smart devices that have communication capability. This has sparked an interest in the development of field networks.

Field networks have been developed with different functional capabilities. Through digital communication, diagnostic capability is available at the source all of the time. All of the fieldbus efforts are aimed at making devices interoperable. Interoperability means that a device can be replaced by another that performs the same function. This is possible through device drivers using a standard device description language (DDL). The DDL supports different blocks, including control at a field device. This means that control can be performed close to the process. This will result in a virtual network that can connect a device to a business application.

The physical networks may be specifically designed for each domain, based on the functional requirements. An information network that integrates business computers will require higher throughput and noncritical information movement, whereas a fieldbus will be more robust and secure and provide data synchronization features. The availability of such network capabilities will change traditional control from DCSs to fieldbus device based control. The higher-level systems will redefine their roles and may evolve to handle information management and advanced control functions.

Alignment with Business Processes

With increased global competition and shorter times to market, business rules change frequently. The information integration requirements between business systems and factory floors will be changing. Additionally, depending on the nature of the business, process automation systems may be required to have the flexibility to change products in little or no turnaround time. This translates to the need for aligning the automation strategy closely to the business strategy.

Software As the Key

The increased capabilities in systems today can provide more flexibility, which is closer to manual operation (see Chapter 3). Software now accounts for 60–70% of the life-cycle costs in an automation system. There will be increasing reliance on software to continue to provide the modularity, scalability, and flexibility.

Need for Scaleable Solutions

The processing industry is moving towards building flexible processes, in which capacity is a commodity. Big plants eventually will give way to smaller, more flexible facilities. The newer plants will need the capability to scale processes very quickly. Automation systems need to be scaleable in order to allow this to happen.

Summary

IRENE: "Wow! This is more complicated than I thought."

ALLEN: "It doesn't need to be that bad. The main thing is to assure that your company has a consistent architectural approach. The control industry is becoming much more open, and using many more common tools. This will, by definition, make these decisions much easier. The more important issue is based on how you will use the system you choose, not which system you choose."

Detailed Design

11

Process Communication

ALLEN: "Previously, we discussed choosing a system. Now let's talk about how to use that system. Earlier, we defined ten key elements of successful automation. One of those elements was 'make your process communicate.'"

LARRY: "What exactly do you mean by that?"

IRENE: "To me, communication requires a successful transfer of knowledge. I know that many of the control systems that are currently installed in our plants rarely transfer knowledge automatically. Usually, someone has to go hunting through tons of information to find what they want."

ALLEN: "That's exactly what I mean by communication."

The Role of Technology in Communication

Technology has revolutionized the way we approach work. It has played a major role in moving manufacturing away from labor-intensive activities through the help of automation. At the same time, it has changed the skills required for operators from task-based to reasoning skills. Many operators have developed a love/hate attitude towards automation, because although it makes their job easier, it also minimizes their role and confuses them by providing information at the wrong time.

Our view of work is being redefined in this age of information technology. Information technology transposes work into an abstract domain of information. Reasoning and thinking are now identified as the most important skills for operators in order to manage newer processes. Do current systems help in this regard? The current design approach to automation systems make it difficult for operators to figure out cause/effect associations.

The automation system should be designed to help decision makers better understand the processes they work with. With the abstraction of work, in which a cause does not directly relate to a visible effect, operators have to visualize the entire process in order to make decisions. Just how do people make decisions?

Human factors is a multidisciplinary science that has developed theories and models of human functioning in working environments, including interaction with systems. Human factors researchers and cognitive psychologists have been developing theories on human behavior for a long time. They represent human behavior using various models. The models help them to compare behaviors, test hypotheses, and make inferences that can aid product design. They have found that systems should be designed based on human behavior. This allows the full potential of automation to be realized.

The SRK Model

The most commonly accepted model is the SRK (skill, rules, knowledge) framework (Rasmussen, 1983). This framework defines a three-level hierarchy: skill-based, rule-based, and knowledge-based behavior. Each layer, depending on signals, generates an appropriate behavioral output.

Skill-Based Behavior

The first level, skill-based behavior, is based on simple stimulus responses. These responses require little conscious activity. An example of a skill-based behavior in process automation is single-loop control, for example, a tank level that has to be maintained at a constant level. When a process is in steady state, and the operator has to control the level in manual mode, all he or she needs to know is the level. If the level drops below the desired value, the operator can quickly infer that action needs to be taken so as to adjust inlet flow. The operator can then execute a specific action to open the inlet valve wider.

Rule-Based Behavior

The next level of behavior is rule-based behavior. This behavior associates input with a task or system state, then selects and executes an action sequence through the utilization of signs. The selection of the action sequence or procedure is based on rules. Let's look at the example of a tank level again. If the level does not return to normal in 15 minutes and the valve is fully open, then something is out of control, and the outlet pump should be shut down. The operator initiating the shutdown procedure uses rule-based behavior to decide this.

Knowledge-Based Behavior

The highest level of behavior is knowledge-based behavior. This behavior is governed by analytical processes and knowledge. When predetermined rules are not applicable, one must use goals and objectives in order to determine a proper response. Once again in the example of the tank level, if the goal of the process is to make product without interruptions, and level is below the limit, the operator may check on the outlet flow rate, and adjust it to an acceptable rate to keep the process running. The operator may then adjust all the downstream process variables accordingly. Someone else may be evaluating why the inlet flow, as reported by the system, does not seem to match reality. In this situation, the operator is responding to an unpredictable event, while demonstrating problem-solving behavior. As a result of this experience, the acquired knowledge may be codified for future use.

Applying the SRK Model to Automation

Process automation systems today already have the capability to handle skill- and rule-based behavior. Automatic controllers and state-based advanced control methods have elevated the role of the operator to knowledge-based behavior where important decisions must be made. Thus, the role of communicating relevant information becomes very important.

Current Situation

Alarm management systems and user interfaces have been the primary interface to the process. Most systems provide more information than is really necessary, with displays proliferating to unmanageable levels. There are instances where operators spend considerable time searching for information in order to intuitively monitor some relation between process variables. True, the automation takes care of most of the procedural tasks, but the information is provided after the system has already responded. If there is an exception that requires operator attention, it is buried in this overload.

Careful thought should be given to the user interface and alarm management in order to keep pace with the changing roles of the user. Information needs and the types of users must be considered before design starts. All data collected should have a defined need. The creation, use, and destruction of the data should then be managed.

Communication versus Information

There is a common misconception that more information is always better. The fact is, in most cases, more meaningless information is thrown at operators than useful information. When a process upset occurs, the operator is usually bombarded with alarms. In reality, the only important alarms are the first one or few that indicate what caused the upset. The rest are simply noise. How can we expect operators to make intelligent decisions, when very little of the information that they receive is useful? Think about taking a test. Nothing confuses students more than giving them too much information.

By modularizing the process and treating each module as an object, you can better manage information. Each module becomes responsible for communicating with the external world (i.e., other modules, the operator, and higher-level systems). These modules should also be responsible to not create meaningless information. For example, a pump CM should disable its low flow alarm as it turns off the motor. Another example is a temperature controller. Whenever the controller is not being used, the deviation alarms should be disabled.

Filtering

One method to minimize the information being sent to the operator is filtering. By filtering the information that the operator receives, you

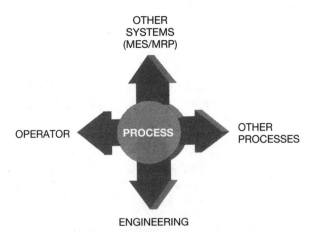

Figure 11.1 Process communication.

can ensure that the information that is received is important. This will also increase the probability that the operator will receive the communication (the knowledge).

Tunneling

When the operator needs more information, they can tunnel into the system to find the information they want. Tunneling is the method by which the operator chooses increasing levels of granularity (and information) as they need it.

Summary

IRENE: "So in other words, we need to look at the information flow, to determine where the information is needed?"

ALLEN: "That's one part of it. The main thing is to manage the information at the modular level."

STEVE: "One area where the process isn't communicating is on alarms. Have you been in a control room lately? There are so many alarms that none of them mean a thing."

ALLEN: "That's a very good point Steve. Let's talk about alarm management next."

12

Alarm Management

OWEN: "I hope that you're not going to bury us with more alarms in this process."

IRENE: "Owen, don't you mean *we*? You're a part of this effort."

OWEN: "Well, we've gotten to this point, and we've barely talked about alarms."

Alarm Definition

One of the definitions of an alarm is information on an event intended to warn of approaching danger. Alarms are intended to attract attention. An alarm could be a stimulus or response for the receiver, however, most control systems treat alarms as stimuli. Alarms serve the purpose of drawing attention to exception conditions occurring within the system. This information is conveyed to the operator in one or more ways: audible horn, graphical display, alarm list, pager, and/or printer message.

Recent regulations have placed increased importance on alarming. Thus, there is a need to understand the principles of alarm design and their shortcomings. The operator interacts with the process in a dynamic setting. Alarms are one of the mechanisms that help an operator to monitor a process. An operator may also discover something is amiss while routinely checking the process. Operators may be under pressure, having to handle multiple tasks, and deal with time-varying data. When the operator is in a problem-solving mode, there are three levels:

- Level 1: recognition and classification
- Level 2: planning

- Level 3: evaluation and monitoring

In the first level, the context of the problem is identified. Once the problem is identified, the operator must decide whether to deploy an existing procedure or to resolve the situation in a new way (level 2). The actual problem is solved at the last level. This process, however, does not typically happen in sequential order. The operator may actually handle all of these things simultaneously (Rouse, 1983).

Alarms have traditionally been associated with control objects. Control was decentralized until relatively recently (e.g., panel-mounted displays). The alarms for each device were generated and presented locally (at the device). DCS systems centralized monitoring functions. With them, the alarms were presented on centrally located display units, usually in a control room. The alarms were presented sequentially, based on the theory that operators manage the process by exception (Dallimonti, 1972). Evaluation of operator's performance of their tasks (Zwaga, 1992) suggests that operators do not perform the tasks by exception. They believe that their function is to prevent alarms, rather than react to them. As process complexity has increased, the number of alarms has proliferated, inundating the operator. Often, due to poor design, the operator has to prioritize alarms based on the state of the process.

For alarms to be meaningful, they need to be managed properly. Alarm management helps to identify, present, and annunciate alarms. Alarms can also be used as a first line of defense to prevent hazardous conditions by presenting the process operator with advance information. Safety instrumented systems can provide the critical alarm information about an upset that led to a shutdown. Automation can integrate these alarms.

Alarm Types

Many control objects have configurable alarms (e.g., a PID block). These alarms can be relative or absolute (see Figure 12.1). Absolute alarms compare the measured parameter to a specified alarm limit. Relative alarms are generated when the measured parameter is referenced to another value or the rate of change. Rate of change alarms are used in applications where the rate at which a variable is controlled or monitored is important (e.g., surge control for compressors or temperature for exothermic reactors).

Discrete control objects, such as valves, can generate mismatch alarms when command and feedback states do not match. If alarms can communicate information about a potentially dangerous situation, they can enhance safety. Often, the deluge of nuisance alarms

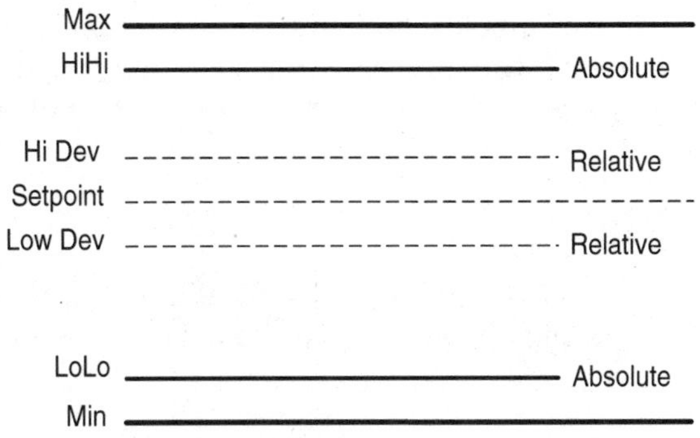

Figure 12.1 Alarm types.

makes the process of discerning critical alarms difficult. This eventually leads to apathy.

Alarms can also be used as diagnostic aids for analyzing process upsets. Advanced control strategies or intelligent systems can help in diagnosing a problem and proposing a solution. These systems have inferential logic or rule sets that are evaluated in the context of the process state. Not all processes can afford to employ intelligent systems, but alarms can at least be made more meaningful.

Alarm Design

Alarms should be separated based on these filter criteria: priority, grouping, and context (Shaw, 1985).

Priority

Alarms should be prioritized based on importance. This distinguishes one alarm from another, so that the operator can respond to a higher-priority alarm first.

Grouping

Alarms should be grouped based on physical location or logical group. This grouping can help operators focus their attention on a specific area.

Context

The process should provide information on when the alarm occurred. The time of the event may be important to the operator. The relevance of an alarm can quickly be identified based on how it relates to the current state of the process. For example, when a unit is being steamed in place, and is out of production, a bad pH alarm may not be relevant.

Suppression

Alarms can also be managed by disabling the annunciation of the alarm. This is done by automation based on the state of the process.

Alarm Management Tools

Most control system vendors have provided tools for alarm management in the system. The standard features for alarm configuration typically include:

- Prioritization
- Grouping
- Ability to enable or disable a specific alarm

If specific alarm reduction techniques or smart alarming capabilities are required, the user needs to configure the alarm management scheme. With the increasing need to find reliable alarm systems, several options are now available:

- Off-the-shelf packages
- Third-party add-ons
- Customized solutions

Nuisance Alarms

Nuisance alarms are alarms that require the operator's attention but do not provide meaningful information. Assume that there is a plant with five units, each having 10 single-loop controllers and 100 discrete devices. If an average of five alarm types are configured for each device, the number of alarms alone is a staggering 2750. To err on the side of safety, and to get consistency in configuration, more alarms are added.

Imagine the operator having to acknowledge up to five alarms in a minute. Monitoring the process while being flooded with alarms quickly overwhelms the operator. A critical alarm could be missed. During engineering, alarm limits are set, but during actual operation, the alarms could go off frequently if the alarm limits are close to the operating range. Studies suggest that nuisance alarms account for 30 to 75% of all alarms (Zwaga, 1992). Initially, this simply distracts the operator, but eventually he or she becomes indifferent to them. Too much information is one of the major causes of failure to detect an abnormal situation (Moray, 1980).

Techniques for Alarm Management

Alarm management provides methods for identifying an alarm and its importance. The primary objective of alarm management is to provide better identification of alarms and reduce the number of alarms. Suppression of alarms is meaningful if diagnostic information is available to the operators. Diagnostic information could include on-line help displays and information to figure out what could have caused an alarm.

Alarms that are less meaningful, alarms for non-operational equipment, or alarms generated during shutdown/start-up should be suppressed. In a case where one alarm triggers several others, identify a first-out critical alarm and store the rest, but don't annunciate them to the operator. Alarms should be segregated based on priority, so that the most important alarms will receive the operator's attention.

There are several techniques available to meet the objectives outlined above: prioritization, suppression, filtering, and smart alarming.

Prioritization

The criteria for prioritization varies from plant to plant. Often, priorities are assigned based on the type of actions required, such as adjustment of control actions. This subjective assignment sometimes defeats the purpose of priorities. Priorities should be assigned during process design. Filtering techniques should be used to further clarify the alarms. Some recommended priorities follow (see Table 12.1).

Priority 1
Alarm conditions that affect safety (personnel, equipment, and process, in that order) should be assigned priority 1. These alarms require the operator to take immediate corrective action. The alarms should be configured in such a way that they need to be acknowledged on a case by case basis. The cause of a priority 1 alarm needs to be investigated if it occurs again. The alarm event and return to normal event should both be printed to a log.

Priority 2
Alarm conditions that are not critical, but still important, should be assigned priority 2. These alarms can serve as a warning that the

TABLE 12.1 Alarm Priority

Level	Acknowledge	Audible	Return to Normal	Log	Example
1 (Critical)	One by one	Yes (Tone1)	Yes	Yes	HiHi, LoLo
2 (Warning)	More than one	Yes (Tone2)	Yes	Yes	Hi, Lo, HiDev
3 (Attention)	Not required	Maybe	No	No	Mismatch
4 (Maintenance)	Not required	No	No	No	Calibration due

process could get to a critical condition leading to a disruption if unattended. These alarms could be acknowledged as a group and printed to a log. Sometimes, they can be used as an indication of suspect product quality.

Priority 3

Alarm conditions that the operator needs to know, but do not impact the process can be grouped into priority 3. Priority 3 is assigned to alarms that have corrective action built in. An example is a valve mismatch due to limit switch failure. The procedural element may employ a workaround for a limit switch failure or change the state of the valve. These alarms need not be audible. If they are, they should be automatically suppressed after a set time.

Priority 4

Alarm conditions relating to equipment, which the operator needs to be aware of, but that do not currently require the operator's attention, are assigned priority 4. These alarms may be logged separately. They may include such messages as "calibration is due" or "fail over equipment selection."

Suppression

This technique suppresses excessive alarms that tend to confuse the operator. If one event activates multiple alarms, and if alarms are propagated to lower levels, then suppression should be used to highlight the first alarm. Alarm suppression prevents notification of the alarm. Conditions leading to the suppression may be sensed or inferred states of variables, including measurements, and the presence of other alarms, or even acknowledgment of the problem by the operator. The state of the process based on the automation can also be used to suppress alarms. Suppression is appropriate when an alarm does not represent a dangerous situation, and is a symptom of a problem that can be deduced from the remaining active alarms. The suppression of alarms should be logged in the same way as process events.

Filtering

Filtering is the process of removing less-important alarms. Alarms that precede the occurrence of another are called precursors. Example, when a Hi Pressure alarm is active and then followed by a HiHi Pressure alarm, the Hi Pressure alarm is the precursor to the latter. Simple logic function can be used to emphasize the precursor alarm. The precursor can be qualified further, depending on the operator's needs.

Missing events may also be meaningful. Filter functions can be used to check for the occurrence of another alarm or event. If the required event does not occur, then a warning could be provided. The event could be a change in process state. This is useful for diagnostics.

Often, an alarm condition will produce other secondary alarms as a consequence. This is sometimes referred to as the "first-out" alarm. Using filters, the subsequent alarms can be deemphasized. Suppression techniques could also be used to disable them.

Smart Alarming

Smart alarming methods help in the task of analyzing an alarm. If an operator wants to know all the possible causes for an alarm based on the state of the process, then more elaborate logic is required. For this, a causal tree is constructed, based on established relationships. Some alarms are the cause of other alarms.

Testing Methods for Alarms

All critical alarms must be tested. The testing should include appropriate alarm response actions. If suppression techniques are used, the methods should be validated first. There should be a clear understanding of suppression methods among the people who design and operate the process. The design effort for creating such a scheme is time-consuming, but if the design and implementation are not validated, problems could result. As a rule of thumb, Priority 1 alarms should never be suppressed.

Alarm Presentation Methods

Alarms can be presented to people in a variety of ways. To convey the state of an alarm when the information is abstract, at least two distinct methods are required: visual and audible cue.

Bells! Whistles!

In a plant, the operator may be responsible for carrying out other activities in addition to running the process. Audible warnings are used primarily to direct the attention of the operator to an alarm. Audible warnings can include bells, whistles, buzzers, horns, and special tones. The primary function of audible warning devices is to attract attention, not to distract. Remember, too many alarms could distract the operator. Audible warnings should use distinct tones that convey the appropriate sense of urgency for the priority of the alarm. Research on the perception of urgency suggests that a burst of syncopated pulses can be used to construct a warning signal for a high-priority alarm (Edworthy et al., 1992).

Displays

Alarms are presented to operators in an alarm display. Alarm displays in traditional DCS systems use a scroll list that updates the latest alarm at the top or the bottom of the display. Given the sequential nature of this alarm display, it is difficult for operators to group alarms. Visual abstraction of an alarm group on an overview display or an annunciator grid is very helpful if it is visible to all operators from a distance. This will help the operator to immediately find the location of the alarm. Flexible filters can help the operator focus on a particular area of interest.

Some alarm management systems have added radio paging, wireless paging, and voice messaging capabilities. However, sending an alarm to a paging device should not be used as the primary form of alarm presentation. Some alarm systems also allow remote acknowledgment.

Process alarms should always be kept separate from evacuation alarms. Evacuation alarms should have different sounds and systems from process alarms.

Advanced Alarm Management

Decision support systems help the operator to diagnose and recover from faults. Decision support systems are becoming important for industries that require a high margin of safety. These systems help decision makers recognize faults and then organize them as root-cause, symptom, or effect.

Tips on Setting Alarm Limits

Since nuisance alarms account for about 40% of all alarms, you should design the alarm management system to minimize them. Before deploying any alarm management technique outlined above, attempt to minimize nuisance alarms by using one of the following techniques:

- For analog signals, minimize the noise by filtering the input signal. Too much filtering may negate the characteristic of the measurement.
- Select deviation limits appropriately based on the capability of the process. Never select a deviation limit tighter than the accuracy of a device.
- For discrete devices, use a time delay on the input or a use a contact-debouncing filter.
- Use absolute alarms for safety (Priority 1) alarms.
- Use relative alarms such as HiDev and LoDev for process alarms.
- Define deadbands for measurement values during design. Anything (or anyone) using this information (controllers, alarm objects) will get an update if the value changes beyond the deadband. This will help to minimize nuisance alarms.
- Disable alarms whenever the signal quality is bad.
- Manage alarm limits using sequential control.

Example

OWEN: "This is all good talk, but where are we going to do it?"

IRENE: "A good place to start is in the analyzer. We have a low flow alarm that should be disabled when we aren't sampling a tank."

ALLEN: "Good point, Irene. How would that look in an operational specification?"

IRENE: "Something like this" (see Figure 12.2).

Summary

OWEN: "The operators will love anything that *we* can do to minimize the number of alarms."

ROBERT: "Whatever we do, we must maintain a safe process."

IRENE: "Allen, won't things such as fieldbus also affect alarms?"

ALLEN: "With the evolution of intelligent devices, alarm management has become more important than ever. An alert object is a new type of control. The object will poll devices, capture the data, and time stamp this information. If the alarms become true objects, some

STATE	RUNNING			STOPPING	RESTARTING	ABORTING	HOLDING
STEP	S1	S2	S3	S1	S1	S1	S1
REQUESTS							
INTERNAL MODULES							
Sample Pump CM	Start		Stop				
ACQUIRED MODULES							
Sample valve to Reactor 1 CM	***		Close	Close		Close	Close
Sample valve to Reactor 2 CM	***		Close	Close		Close	Close
ALARMS							
Low Flow	Disable	Enable	Disable	Disable	Enable	Disable	Disable
TRANSITIONS							
Unit ready to transfer and Fow > FSL	T1						
2 Minutes		T2					
Pump off			T3	Complete		Complete	Complete
MESSAGES							
Waiting to initiate sample	151						
Sampling		152					
Analysis complete			159				
PARAMETERS							
P1 Test #							
REPORTS							

*** One of the valves open based on which tank is being sampled, the other closes.

PERMISSIVES AND INTERLOCKS

CONDITION	TYPE
Flow below FSL after step S2	I

P: PERMISSIVE; I: INTERLOCK

Figure 12.2 Operational specification for alarm management.

of the alarm management methods can be embedded in the object itself."

IRENE: "It seems like alarming is a critical part of the operator interface."

ALLEN: "It certainly is, and operator interface is what we are going to discuss next."

13

User Interface

IRENE: "Now that we've started the project, we should start on the screen designs and menus."

ALLEN: "Does that mean user interface? Can you explain how you create the operator interface?"

IRENE: "I don't know what you mean by user interface, but I can explain how we build our operator interfaces."

ALLEN: "You're referring to screens, operator interfaces, and menus in the same breath. Let me explain some differences:

- A user interface is the primary interface for human–computer interaction. The user interface can have displays, menus, and icons.
- A menu is a list of options presented to a user."

IRENE: "Then I must be referring to the user interface. We use the graphic development tools that the vendor provides. We get input from the operating team on how the displays should look while we are building them. Then we build and test them using the tools that are available."

ALLEN: "How do you define the level of detail, the navigation among displays, and the overlays?"

IRENE: "We have to use the standard options available from the vendor. The choices are limited. I really can't do what I would like to."

STEVE: "I've been in many start ups. Even though some of the automated processes used the same control system, each one was different."

IRENE: "We helped put together the displays we liked in the last project. It worked out well. Every graphic screen that interacted with the process was checked in detail."

STEVE: "What I'm trying to say is that the animation schemes didn't match. The static graphics diverted my attention from the actual process events. The color schemes meant different things for each plant."

IRENE: "I would agree with that. But we provide what the plant asks for."

ALLEN: "Have you consistently defined what the user interface should be for all your plants?"

LARRY: "I don't think that's necessary. We have plants of different vintage ranging from ones using pneumatic instruments to those with state-of-the art systems. Each one uses different levels of technology and I think the user interface varies with the level of technology."

ALLEN: "True, there are different levels of technology, but if you need consistency among your operations, you need to define some principles of user interface that are common to all operations."

LARRY: "There was a committee that came up with recommendations. They even published an internal standard, which we're supposed to follow."

IRENE: "That was a panel layout standard published ten years ago. We don't apply those standards to any of our new projects."

VICTOR: "Why do I need standards for user interface? We already have standards for everything. If you build the interface intuitively, it should be easy for the operators to work with."

ALLEN: "There are major benefits in standardizing user interfaces. The most important is consistency. When solutions are built based on applying these standards, it helps to reduce training time. The operators can work with different processes. They don't have to undergo system-specific training, but can focus on learning the process. The support staff can support multiple systems without having to remember the different display hierarchies. A consistent design also ensures that inadvertent mistakes aren't made. The second major benefit is the reduction in engineering time. You don't have to reinvent the wheel each time you develop a user interface. A third benefit is the ease of making changes. You don't have to rely on a developer to make these changes. The user interface standard is ideal for a toolkit."

IRENE: "The toolkit has to be specific for a particular system. We haven't selected our system yet."

ALLEN: "The automation plan addressed the selection of the vendor and the system. A toolkit for user interface can be a good example of the working standard implementation."

OWEN: "Why are we talking about standards for user interfaces when we don't have one that works?"

VICTOR: "Why don't we use our standards. We spent a lot of effort on them."

ALLEN: "Let's try to understand why the standard is not being used."

IRENE: "Well, as I understand it, it was about panel layout. In the last edition, the standard made references to new systems. It said motif would be the windows standard."

ALLEN: "I studied the standard. I see several reasons why that standard did not succeed. Primarily, the standard didn't anticipate future interface trends and needs. The standard also mixed procedures, guidelines, and normative references. You can't have an enforceable standard if it's not meant to be enforced. In looking at various plants, it's obvious that that standard was not enforced."

VICTOR: "We have a consensus process for building standards. You mean to say that a consensus standard may become useless?"

LARRY: "We make sure that all of our engineers have a copy of the standards. We also make sure that they are aware of them. I know engineers who use them."

ALLEN: "What we need is follow-through on design standards like these. The follow-through activities should involve dissemination, training, tools, enforcement, and support."

VICTOR: "What should a user interface standard be like? How do you build one?"

ALLEN: "There are several steps you need to follow to build a user interface standard:

- Get the right people involved
- Get information
- Prototype and write the standard; get consensus
- Final review
- Follow-through

For a user interface standard, get the right people involved. This should include operators, a human factors expert, an automation engineer, and a process safety engineer."

IRENE: "Just what is a human factors expert?"

ALLEN: "Human Factors experts relate visual perception to cognitive psychology. By doing this, they can help in developing interfaces to computer systems. For example, the software industry uses human factor experts to design and test user interfaces."

LARRY: "I do not understand why a safety engineer should be involved."

ALLEN: "Safety engineers can help to understand alarm grouping and systems interactions when processes are not safe."

ALLEN: "The next step is to get information. This process should provide the necessary details on technologies to deploy standards and should also list the user interface standardization that exists. Before the standard is written, the scope and purpose should be clearly defined, even though the need for the standard is clear."

ALLEN: "The next step is to write the standard. Do the prototype for the standard at the same time. The prototype helps as an effective discussion aid. Use a specific vendor system or general drawing package to develop the prototype. Then subject the standard to a review. The review audience should include people who interact with the process at all levels. After the review, publish the standard. The guidelines and principles used in deriving the standards are for the people who will build these systems."

ALLEN: "The most critical step is follow-through, however. You can't publish a standard and then forget it. The best standards are the ones that are flexible and alive. A standard can live only when people follow it."

IRENE: "You mentioned guidelines. Earlier you talked about technology standards. Where does this standard fit in?"

ALLEN: "The user interface standard is a design standard."

IRENE: "Can you help me understand what the guidelines are?"

OWEN: "Sure, this involves the color of a pump, the color of an alarm, etc."

STEVE: "I'd like to see that color schemes are the same for all systems."

ALLEN: "You're right, but the guidelines should be derived from fundamental principles. The user interface allows the operator to communicate with the process. The design should be simple, elegant, and provide the user with the right information at the right time."

IRENE: "Right information is a catchall phrase."

OWEN: "We're used to working with blocks and an overview graphic. We get the information we need, but sometimes it's too much information or it's not relevant."

ALLEN: "Good user interfaces provide the right amount of information on the health of the process. When there's an upset, the interface should have retrieval and navigation aids to get all of the information the operator asks for at whatever level of detail. This approach helps users move towards adopting a process management view. We discussed that earlier when talking about process communications."

User Interface Overview

User interface is the mechanism by which people interact with a system(s). The interaction may be passive, as simple as monitoring the system, or active, such as controlling the process. User interface includes all of the devices that help to monitor, control, annunciate, and supervise a process. The user interface can provide information in several ways, using text (inclusion or omission), pictures (graphics), sound (alarms), and color.

Most systems provide tools to build user interfaces. Some of the tools available on control systems include faceplates and prebuilt display objects (widgets). These tools could be used to build displays, by facilitating the development of the interactive environment. The environment can use windows, menus, device interaction (faceplates), and message areas. Different levels of customization are required to build a meaningful interface. Some plants may have displays based on faceplates alone. In building the user interface, remember the different users of the system and tailor it to their specific use.

The user interface standard should take into account the profile of its users. The profile includes age, physical abilities, education, cultural or ethnic background, motivation, goals, and personality. Remember, the users could be novices in the system, but knowledgeable about the process.

An operator may have to know the status of the process and provide inputs to control it. A manager may want an overview of the process with some performance indicators. A maintenance technician may want to have access to a device and its parameters to troubleshoot a device. A system administrator will want system diagnostics. The user interface standard may focus only on the custom displays through which the interaction to the process takes place.

The most popular technology standard for user interface in Process automation is now the windows system. This style offers some distinct benefits:

- Different types of information can be displayed simultaneously.
- Many tasks are available through a pull-down menu.
- The systems are more user-friendly. These techniques reduce the amount of typing, thereby making interaction easier.

The benefits of a user interface can be fully realized if properly designed. A user interface standard helps to accomplish this. Good practices and guidelines are considered before a standard is created. The following section outlines some of the guidelines.

Design Guidelines for Building User Interfaces.

There are several principles that should guide the user interface design standardization. The primary guidelines are common to all types of user interfaces:

- Consistency
- Feedback
- Verification
- Organization
- Choice of elements

Consistency

Use a consistent format in the menu structure, display elements (graphics), and interaction with functions of the system (mode and state changes, for example). If someone needs to change a parameter, do not provide different ways of accomplishing it in separate places.

Feedback

Provide feedback to ensure communication.

Verification

Any critical entries to the system should be verified. For example, a mode change request should require two keystrokes. This will also prevent accidental entries.

Organization

Organization of the elements on the interface is very important. The organization applies to screen layout and functions. The screen layout should be such that users will easily learn where to look. The functions should be organized in an intuitive manner.

Choice of Elements

Several elements are available. The selection of these elements should be carefully thought out. For example, if a combination of touch screen and point and click devices are used, the touch screens need to have a verification for data entry. If the purpose of the user interface is only to provide informational displays, then complex menu structures can be avoided.

Detailed Design Checklist

The following checklist is a guide for a good user interface design. The checklist is based on human factors research.

Screen Layout

Screen layout should follow established user interface standards. Natural eye movement always starts at the top left and then moves right and down. The screen layout should match eye movement. This will give a clear sequence to the screen. When using text on labels, always left-justify them.

When using pop-up windows, size them so that the pop-ups do not cover important process information displayed in the primary window. For example, when using a pop-up data entry screen to change a setpoint, the pop-up shouldn't cover the graphics area of the screen that conveys the status of the process.

Icons and Buttons

Use icons and buttons to implement specific commands. Large buttons are recommended, since they are easier to select. Button animation should be consistent. Whenever possible, provide feedback showing that a command is being processed. For example, ghost the text or change its color. If there are only two states for a command button, consider creating one button, rather than two. For toggle commands, consider changing the text inside the button. Icons are good for functions or applications unrelated to the primary task. Application Icons need to be labeled. An example of an application that might have an icon is a report writer.

Menus

Menus are good for infrequent commands or secondary users. You should group the choices in the menu by function. If the menu is a pull-down menu, group the menu picks by function. Functional grouping helps users decide where to look for an activity pick. A pull-down menu should have frequently used activity picks at the top. Derive the menu structure after careful thought. Try to sequence items in a pull-down menu. Use grouping with lines or indents for subgrouping of functions.

Pop-ups

When using pop-up or dialog boxes for interacting with the user, keep the dialog in the context of what needs to be done. Try to group fields by function or subfunction. The dialog box should not be cluttered. The dialog should be self-explanatory to accomplish the function at hand.

Color

People like the use of color because it provides visual stimulation and it helps in searching. Six colors are ideal to convey information. More than six colors makes identification difficult.

There are several things to remember when using color. When choosing a color scheme, follow the color stereotypes that everyone relates to. Choosing a favorite color and overdoing it can be annoying to operators and make the interface difficult to use. There are several problems in using color. Do not rely only on color to convey information, since research suggests that everyone is colorblind in low light. Screen problems such as contrast, aberration, and fuzziness can also occur. Pure red and blue text has the tendency to create aberrations. The background color for primary windows should not be jarring. Use a soft color, like gray.

Animation

Animation is necessary to provide tactile feedback on the state of the process. For example, to depict the changing level of a tank, color and vertical motion can be used. When an alarm condition occurs, color change can be used to indicate the condition. This provides the user with an absolute and relative position. The animation should be carefully thought out. The user can get overwhelmed when too many fields blink or change colors at the same time. Varying rates of blinking will complicate a display. Animation can also be used to show a command in progress.

Text

Use simple words in text. Only use acronyms (or shortened forms of words) when all users are familiar with them (e.g., auto or man). Wording should be as precise as possible. Use lowercase because uppercase takes longer to read. For commands, use uppercase.

Messages

Messages serve the purpose of conveying information to the user. They may be used to prompt the user to perform some function or display an error or status message. Avoid unnecessary words in messages. Use active verbs and sentences. If a single window is used for all messages, use visual or audible cues. This will prompt the user's attention. A change in color for messages is recommended, based on the type of message. Provide a delay time (0.5 seconds) for generating data input error messages.

Graphics

Graphics help users to better perceive the process. The graphics layout should be optimized to reduce clutter, yet present meaningful information. The attributes that are not relevant for the operation of a process need not be depicted on the graphic. For example, the location of nozzles or platforms for a column are not required.

Graphics should be built from a standard library. The library should be part of the user interface standard and may contain widgets or just static graphics. The widgets should have preconfigured animation states and placeholders defined.

Interaction

Let the user control the interactive flow. The system should not force the user to interact with the system in a particular way. The use of devices in the operating environment should be clearly understood. For example, a mouse or track ball is a pixel-pointing device. It is suitable for large cursor movements. A mouse interface is difficult to use in an area where the operator must stand, since it requires both hand and eye coordination. For frequent operations, function keys or keyboard buttons are ideal. If the user interface has both a keyboard and mouse, avoid forcing users to switch between devices frequently.

Interface Perspective

There are two fundamentally different types of operator interface: process-based and product-based. In any process, we need both views, but the more and better the automation, the more important the product-based view becomes.

Process-based views should start with a high-level overview and allow the operator to tunnel down as needed. They focus on the status of the equipment—Things such as what is running, what valves are open, and what batch is in the unit.

Product-based views show what products are being made, where they are, and what is happening to them. As the operator moves more into a role of managing the process, the product-based view becomes increasingly more important.

Summary

VICTOR: "We committed to an automation plan. One of the activities is the user interface standard. We need to get it done. Can we use our project to showcase the implementation of the standard?"

ALLEN: "Showcasing a project that has implemented standards is an excellent way to get a following. There is one last thing to remember when designing an operator interface. It is only a way to allow the process to communicate with the operator. The screens in and of themselves do not add value. Think of the displays as being a transparent interface into the process."

Programming Techniques

ALLEN: "There are as many different ways to program a computer as there are programmers. Fortunately, there are some standard programming techniques that can allow code to be more portable, understandable, and reusable both within and across platforms."

IRENE: "We just finished talking about object oriented design and programming. That is obviously one of the techniques. What else do you mean?"

ALLEN: "Object oriented techniques do much to enhance the reusability of code. However, there are other ways to accomplish this. Programming should be based on general models and templates rather than being unique to each project. There are also ways to make the code for a project independent of the system on which it will be implemented."

IRENE: "Are you talking about IEC1131-3, and if so, are programs developed for one system really transportable to another system?"

ALLEN: "Yes, I'm talking about IEC1131-3, and no, the code is not necessarily transportable, at least not yet. Although, if the design is done using IEC1131-3 tools, the design could be implemented on another system. In most cases though, the actual program will need to be recreated for the individual system on which the automation will run. Allow me to go through some of the other issues pertinent to programming your automation."

Model-Based versus Custom

Develop templates or models from which your automation is developed. Develop phase templates for standard phases. Develop standard CMs that can be copied and used in most cases (e.g., block valve with one output and a closed limit switch, block valve with one output and both closed and open limit switches). These CMs then have methods that are associated with them that are activated through messages or commands (e.g., open).

Open versus System-Dependent

What does it mean to be "open" versus "system-dependent" or "proprietary"? In the case of the actual automation code, one meaning is that the code is developed in a consistent language (for example using 1131-3). It also means that the specification for the automation (operational specification) is written in a platform-independent manner.

IEC 1131-3: Programming PLCs

IEC 1131 is a standard from the International Electrotechnical Commission for PLCs. Subpart 3 (1131-3) is titled "Programmable controllers—Part 3: Programming languages." According to the standard:

> This part of IEC 1131 specifies the syntax and semantics of a unified suite of programming languages for programmable controllers (PCs). These consist of two textual languages, IL (Instruction List) and ST (Structured Text), and two graphical languages, LD (Ladder Diagram) and FBD (Function Block Diagram).
>
> Sequential Function Chart (SFC) elements are defined for structuring the internal organization of programmable controller programs and function blocks. Also, configuration elements are defined which support the installation of programmable controller programs into programmable controller systems.

Structured Text (ST)

Structured text is closest to traditional computer programming. It is a Pascal- or C-like language utilizing standard programming constructs. Table 14.1 shows some of the standard constructs used by ST. Structured text is typically used for higher-level functions such as supervisory control or complex logic. However, it can be used to perform almost any form of control.

Sequential Function Chart (SFC)

Sequential function charts are used to program in a defined order (sequentially). SFCs consist of steps and transitions. Each step points to another SFC or another form of control (LD, FBD, ST, or IL). Transitions cause the control execution to move between steps. Figure 14.1

TABLE 14.1 Structured Text Constructs

IF *CONDITION* THEN
    ACTION;
ELSEIF *CONDITION* THEN
    ACTION;
ELSE
    ACTION;
END_IF;
	CASE *VARIABLE* OF
	X: ACTION;
	Y: ACTION;
	Z: ACTION;
	ELSE
	    ACTION;
	END_CASE;

FOR X := Y TO Z BY W DO ACTION; END_FOR;	WHILE CONDITION DO ACTION END_WHILE

shows an example of the use of an SFC. The boxes correspond to steps, and the short horizontal lines (dashes) correspond to transitions. The entire chart is processed from the top down and left to right. Multiple, single horizontal lines at the same level correspond to conditional paths (IF ELSEIF ELSE). These conditions are processed from left to right. The first transition to be true causes the control after that transition to be performed. All transitions and paths to the right of the first true transition are ignored. Double horizontal lines correspond to parallel processing. All paths under the parallel horizontal lines are performed simultaneously.

Due to their sequential nature, SFCs are very often used in procedural control. They are an intuitive way to represent the sequence of actions that will be performed by a procedure. There are some major

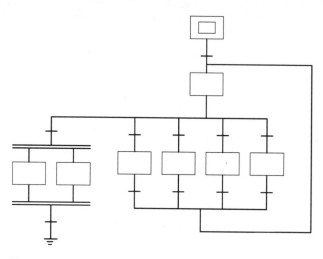

Figure 14.1 Sequential function chart.

constraints that are placed on the execution of the control. For example, the control must follow the path that is drawn in the SFC. There are no jumps, and therefore, if an exception condition necessitates a move from one part of the chart to another, a transition and link must be present

Ladder Diagram (LD)

Ladder diagrams are the traditional forms of programming for programmable logic controllers (PLCs). Ladder logic was originally developed as a replacement for electrical relays in applications such as motor starter circuits. PLCs were a configurable way to implement the same control as inherent in electrical relays. Over time, many functions beyond simple relay replacement were added. In most cases, PLCs have even incorporated forms of analog control and floating point math into their ladder logic functionality (Figure 14.2). Ladder diagrams are best suited for discrete control and interlocking. With the addition of other programming environments, the need to perform floating point math and analog control using ladder diagrams has lessened.

Function Block Diagram (FBD)

A function block (FB) is a graphical representation of a control algorithm. The actual functionality of the FB is implemented with some internal code. Sometimes, that may be one of the other IEC 1131-3 languages, but quite often it will be an actual function block that came with the control system. Function Blocks are configured by placing a given FB in an FBD. An FB can have inputs and outputs that

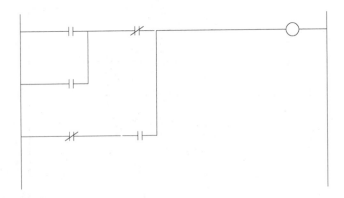

Figure 14.2 Ladder diagram.

are linked either to variables in the control system, constant values, or other another FB. An example of an FBD is shown in Figure 14.3.

Instruction List (IL)

Instruction list programming is similar to assembly language programming. It consists of short pneumonically based operations and associated operands. It can be used for almost any form of control. It is very cryptic and quite difficult for someone, other than the person who wrote the program, to interpret and has yet to find widespread use.

Phase Programming

Phase programming should be based on a clearly defined set of templates. The actual work to be performed during the phase is then plugged into the template for each individual phase. There are three common forms of phase programming: SFC, structured text, or ladder logic. We are going to focus our discussions on SFC and structured text.

Parallel versus Sequential Programming for Phase Logic

When programming phases using SFCs, two primary methods are employed: sequential or state machine (see Figure 14.4). There are advantages and disadvantages to each.

Sequential phase programming is more intuitive and easier to troubleshoot. The SFC shows the order of step execution, and most control systems will show what step is active. SFCs are, on the other hand, more difficult to make changes to. They also have problems with recovery from an exception. Whenever an SFC is started, it must start at the first step and progress only through connected steps and transitions. If, after an exception, you would like to restart at a step other than 1, you must show a transition from the starting point to each

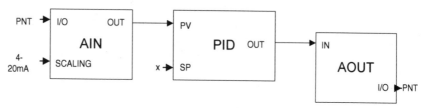

Figure 14.3 Function block.

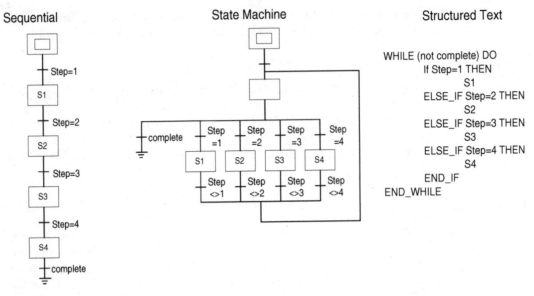

Figure 14.4 State machine programming.

and every step in the SFC (see Figure 14.5). Also, in order to change the order of steps, the SFC must be redrawn.

State-machine-based programming, also shown in Figure 14.4, can be implemented in either SFC or structured text. With this implementation, the phase can move from any step to any step, based on transitions. This allows new steps to be added and/or steps to be resequenced very easily. It also allows recovery from an exception without any changes. One drawback, however, is that this form of programming is less intuitive. It doesn't show the sequential nature of the steps.

Single-Step

Single-step is a powerful mechanism that can be implemented to aid in troubleshooting and checkouts. When a phase is in a single-step mode, the transitions are not processed until the operator issues a resume. When a resume is received, the phase will process the transitions and move to the next step and wait for another resume. This can allow someone to verify the control actions taken by each step within a phase.

On the other hand, if a phase somehow ends up in a single-step mode during normal process operation, serious problems could result. Care should be taken to never allow a phase to get into a single-step

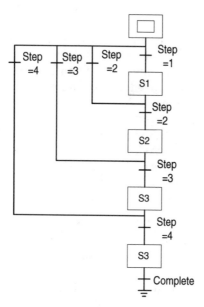

Figure 14.5 Phase recovery.

mode except during troubleshooting and checkouts. This can be accomplished through a security feature or by disabling the single-step
functionality. An example of the code that could be used to implement
single-step is:

```
IF SINGLE_STEP THEN
        PAUSED := TRUE;
END_IF;
```

The paused flag is then used to disable the processing of transitions
(moving between steps) within the phase.

Pause

Pause is a mechanism that can allow the operator to halt execution of
a phase at a safe point. When a phase is in the Paused state, the transitions are not processed until the operator issues a resume. Once a
Resume is received, the phase continues normal operation. An example of the code that could be used to implement pause is:

```
IF Pause THEN
     PAUSED := TRUE;
END_IF;
```

Resume

From either a single-step or a pause, some mechanism must be implemented to resume normal operation. When a resume is received, the phase should move to the next step when in single-step or continue normal operation when paused. An example of the code that could be used to implement Resume is:

```
IF Resume THEN
      PAUSE := False;
      PAUSED := False;
END_IF;
```

Phase State Model

Before designing a phase template or any individual phases, you should first determine a state model for the phase. Most batch management systems specify a phase model. For illustrative purposes we are using the following state model in Figure 14.6.

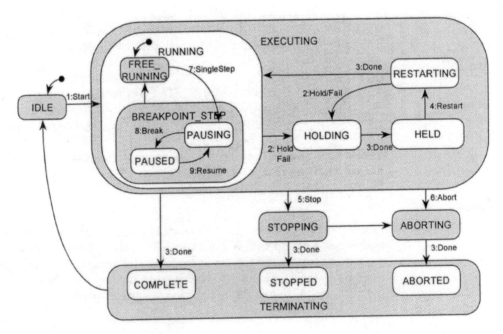

Figure 14.6 Phase state model.

For this phase state model, the following states must be programmed: running, holding, restarting, stopping, and aborting (for a complete description of the phase state model and interface, see Appendix E). The running logic is the normal phase operation. The holding logic is the exception handling logic. The restarting logic handles recovery from an exception. The stopping logic includes the logic to stop phase execution when it is externally terminated. The aborting logic includes the logic to abort the execution of the procedure when received from an operator. Note that in this state model, abort is used to abort the execution of a batch and hold is the exception handling logic.

Phase Synchronization

Most batch management systems have the capability to coordinate the execution of phases. This capability should be used whenever possible. By allowing the batch management system to synchronize phase execution, the amount of peer-to-peer communication required between PLCs is minimized. The amount of work needed to change or add phases in the future is also minimized. For example, by allowing a batch management system to coordinate a XFER_OUT and XFER_IN phase pair, another unit can be added with a XFER_IN, and the original XFER_OUT phase should require little if any change.

Allocation and Arbitration

Whenever possible, allow a higher-level system to perform allocation and arbitration functions. This will minimize programming changes in the future due to new products or changes to upstream or downstream processes.

Treat Modules as Objects

A phase should always treat EMs and CMs as objects. That is to say that it should never act directly on an output. It should send a message to a CM requesting it to take an action. For example, when a phase wants to start a pump, it should send a 'start' message to the pump CM (see Figure 14.7). The pump CM will then take the appropriate actions to start the pump and verify the status of the motor. The phase will then verify that the pump is running by looking at the pump CM's status attribute rather than looking directly at the pump motor status input. In this way, the CM can take all other things that affect the pump into account before setting the status.

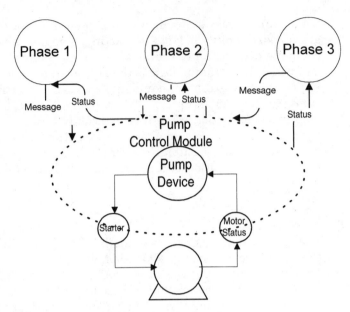

Figure 14.7 Phase to CM communication.

Structured Text Phase Template

The following code is a structure text phase template:

```
IF ABORTING THEN (*TEMPLATE FOR 1 STEP ABORT*)
    IF (STEP_NUMBER = 1) THEN (*PERFORM STEP 1*)
        (******PUT THE WORK FOR ABORT STEP 1 HERE *)

        (******PUT THE ABORT COMPLETE CRITERIA HERE *)
        (*IF CONDITIONX THEN*)
        ABORTING_COMPLETE := TRUE; (*SET THE ABORT COMPLETE FLAG*)
        (*ENDIF*)
    END_IF;
ELSIF STOPPING THEN (*TEMPLATE FOR 2 STEP STOP*)
    IF (STEP_NUMBER = 1) AND (NOT PAUSED) THEN (*PERFORM STEP 1*)
        (******PUT THE WORK FOR STOP STEP 1 HERE *)

        (******PUT THE TRANSITIONS OUT OF STEP 1 HERE*)
        (*IF CONDITIONX THEN*)
```

```
     STEP_NUMBER := 2;
     (*END_IF*)

     (*THIS PAUSES THE PHASE AFTER THIS STEP IF SINGLE STEP IS
ACTIVATED*)
     (******THIS CODE SHOULD BE DELETED IF SINGLE STEP ISN'T
ALLOWED AT THIS POINT*)
     IF SINGLE_STEP THEN
      PAUSED := TRUE;
     END_IF;

   ELSIF (STEP_NUMBER = 2) AND (NOT PAUSED) THEN (*PERFORM STEP 2*)
     (******PUT THE WORK FOR STOP STEP 2 HERE *)

     (******PUT THE STOP COMPLETE CRITERIA HERE *)
     (*IF CONDITIONX THEN*)
     STOPPING_COMPLETE := TRUE; (*SET THE STOP COMPLETE FLAG*)
     (*END_IF*)
   END_IF;
ELSIF HOLDING THEN (*TEMPLATE FOR 2 STEP HOLD*)
   IF (STEP_NUMBER = 1) AND (NOT PAUSED) THEN (*PERFORM STEP 1*)
     (******PUT THE WORK FOR HOLD STEP 1 HERE *)

     (******PUT THE TRANSITIONS OUT OF STEP 1 HERE*)
     (*IF CONDITIONX THEN*)
     STEP_NUMBER := 2;
     (*END_IF*)

   ELSIF (STEP_NUMBER = 2) AND (NOT PAUSED) THEN (*PERFORM STEP 2*)
     (******PUT THE WORK FOR HOLD STEP 2 HERE *)

     (******PUT THE HOLD COMPLETE CRITERIA HERE *)
     (*IF CONDITIONX THEN*)
     HOLDING_COMPLETE := TRUE; (*SET THE HOLD COMPLETE FLAG*)
     (*END_IF*)
   END_IF;
ELSIF RESTSARTING THEN (*TEMPLATE FOR 2 STEP RESTART*)
   IF (STEP_NUMBER = 1) AND (NOT PAUSED) THEN (*PERFORM STEP 1*)
     (******PUT THE WORK FOR RESTART STEP 1 HERE *)

     (******PUT THE TRANSITIONS OUT OF STEP 1 HERE*)
     (*IF CONDITIONX THEN*)
     STEP_NUMBER := 2;
```

```
     (*END_IF*)

     (*THIS PAUSES THE PHASE AFTER THIS STEP IF SINGLE STEP IS
ACTIVATED*)
     (******THIS CODE SHOULD BE DELETED IF SINGLE STEP ISN'T
ALLOWED AT THIS POINT*)
     IF SINGLE_STEP THEN
      PAUSED := TRUE;
     END_IF;

     (******INSERT THIS CODE IF AFTER THIS STEP IS A PAUSE POINT*)
     (*IF PAUSE_COMMAND THEN
      PAUSED := TRUE;
     END_IF;*)

   ELSIF (STEP_NUMBER = 2) AND (NOT PAUSED) THEN (*PERFORM STEP 2*)
     (******PUT THE WORK FOR RESTART STEP 2 HERE *)

     (******PUT THE RESTART COMPLETE CRITERIA HERE *)
     (*IF CONDITIONX THEN*)
     RESTARTING_COMPLETE := TRUE; (*SET THE RESTART COMPLETE
FLAG*)
     (*END_IF*)
   END_IF;
ELSIF RUNNING THEN (*TEMPLATE FOR2 STEP RUN*)
   IF (STEP_NUMBER = 1) AND (NOT PAUSED) THEN (*PERFORM STEP 1*)
     (******PUT THE WORK FOR RUN STEP 1 HERE *)

     (******PUT THE TRANSITIONS OUT OF STEP 1 HERE*)
     (*IF CONDITIONX THEN*)
     STEP_NUMBER := 2;
     (*END_IF*)

     (*THIS PAUSES THE PHASE AFTER THIS STEP IF SINGLE STEP IS
ACTIVATED*)
     (******THIS CODE SHOULD BE DELETED IF SINGLE STEP ISN'T
ALLOWED AT THIS POINT*)
     IF SINGLE_STEP THEN
      PAUSED := TRUE;
     END_IF;

     (******INSERT THIS CODE IF AFTER THIS STEP IS A PAUSE POINT*)
     (*IF PAUSE_COMMAND THEN
      PAUSED := TRUE;
```

```
    END_IF;*)

ELSIF (STEP_NUMBER = 2) AND (NOT PAUSED) THEN (*PERFORM STEP 2*)
    (******PUT THE WORK FOR RUN STEP 2 HERE *)

    (******PUT THE RUN COMPLETE CRITERIA HERE *)
    (*IF CONDITIONX THEN*)
     RUNNING_COMPLETE := TRUE; (*SET THE RUN COMPLETE FLAG*)
    (*END_IF*)
  END_IF;
END_IF;
```

Summary

IRENE: "Now I see how the phases that we defined earlier will be programmed easily into the phase templates that we just developed. I'm starting to see why we spent so much time up-front defining everything."

PATTY: "Allen, is the IEC 1131-3 standard really that usable?"

ALLEN: "Many control systems are starting to use a modified form of some or all of the programming languages."

15

Object Technology

IRENE: "We've had many discussions about programming our control system, but we haven't discussed object-oriented programming. Is there a reason for that?"

ALLEN: "Object-oriented programming, or object technology, is an important part of modern programming practices, but it's just beginning to be applied to the field of process control."

MARK: "I've heard a lot about object-oriented programming, but what exactly is it?"

ALLEN: "Let's look at a brief overview of object technology."

Background of Object Technology

Process automation engineers are faced with many choices when selecting software on how to implement systems. Computer programming has evolved over the years. Thus, several programming methods are available for the configuration of control systems. Three programming techniques that are commonly used for control systems are: unstructured programming, procedural programming, and modular programming.

More recently, object-oriented (object) technology has evolved. Object technology is based on modular programming techniques. The software industry seems to have embraced object technology. This has resulted from several factors:

- Increasing complexity of new applications
- Large time commitment in developing and deploying new software
- Soaring software maintenance costs

Object technology has the potential to simplify the creation of software, reduce development time, and provide more flexibility. Advances in computing technology are making this possible. The increased need for flexible systems is the result of the changing needs of the global business economy.

In the following sections we attempt to demystify object technology and explain its potential impact on automation. We will also explore how the convergence of the S88 models and object technologies can offer an alternative to the conventional approaches to building process automation systems.

Object technology is a new way of thinking about, organizing, and developing software. It includes the following principles, methods, and tools:

- Object-oriented programming languages (OOP)
- Object-oriented analysis and design (OOA-D)
- Databases (OODBMSs)
- User interfaces (OOUIs)
- Operating systems
- Components
- Frameworks
- Object request brokers (ORBs)

What is an Object?

An object is the fundamental basis of object technology. We all think of objects as physical things that we find around us. Cars, toys, bicycles, or humans are all examples of objects. A company could view employees, buildings, process reactors, or recipes as objects.

The software industry has extended this view of an object to also include a piece of software consisting of its own data and private procedures (methods). From our modular manufacturing discussions, an object is a module. It is a self-contained module of code that has a defined appearance (attributes) and interface to the external world (methods).

The *state* of an object is the condition of the object, or a set of circumstances describing the object. The state of the object is part of its data. For example, the state of a pump could be "on" or "off." The state of a reactor could be more complex. However, we can restrict the state to something relevant, such as "running/normal/automatic," since the object is a model of reality.

Procedures, also called methods, perform the actual work of the object. An example of a method is the opening of a block valve when it receives an open message.

Each object has a unique identity. The uniqueness of the object is achieved by the object identifier. Objects are distinguished from each other without comparing their values or their behavior.

What Is Different about Object Technology?

Historically, programs were written so that:

- Data and procedure were different
- Work was done when the procedure operated on the data
- The result was modifications to the input data or new data

Software written in this manner can quickly become complicated, because data produced from one procedure will be used as input to other procedures. Furthermore, it is not possible to foresee all possible consumers for this data. With objects, however, the basic organizing scheme is different in that it encapsulates both data and methods.

Object Classification

Objects can be classified into three types:

- Functional objects
- Imperative objects
- Active objects

Functional Objects

A functional object has no state and no identity. It has an object-like interface.

Imperative Objects

Imperative objects are also referred to as *passive* objects. The traditional definition of objects applies to them, but they remain passive or static until something outside the object requests a change of state. This activation is done by sending a *message* to them.

A message is sent from one object to another. There are three components to a message:

- Receptor. The receptor specifies the receiving object.

- Selector. The selector specifies the appropriate method to invoke as a result of the message.
- Parameters. Parameters are function arguments or operands, which the method associated with the specific message needs.

The receiving object responds by executing a corresponding method. The receiver responds back to the sender by returning a value. An example of a message is:

aUnit123 makeBatch: aBatchID usingRecipe: Coco

Note that aUnit123 is the receiver object and has methods called makeBatch and usingRecipe. The result could be aBatchIDReport object.

Active Objects

Active objects may be executing when a message arrives. In simple terms, active objects can alter their own state.

Classes

A class is a template for creating objects. In object-oriented programming, if you have to create an object every time you need one, then the benefits of reusability may not be realized, especially if the objects need similar characteristics. A class is a template from which objects may be created. It contains a definition of the state descriptors and methods for the objects.

Classes serve three purposes:

- Structural definition. A class defines the names of attributes and methods of an object belonging to this class.
- Production of new objects. The class is used as a clone factory to create new instances. The mechanism of deriving new objects from a class definition is called instantiation.
- Classes bring order to the world of objects. The class is used to administer a set of instances of the class.

Each instance of a class has the same structure as all other instances of that class. Each object shares its instance methods with all other instances of its class. Methods are the shared parts of an object.

A metaclass is a class whose instances are themselves classes. When the instance creation mechanism of the metaclass creates an instance, the instance created will be a class.

A parameterized class is a template for a class wherein specific items have been identified as being required to create nonparametrized classes based on the template. In other words, the instance creation mechanism is used to create a nonparameterized class rather than an instance. Instances can then be generated from its instantiation mechanism. For example, a parameteized class called "Controller" might have a specific item named "ControllerType," a class name. The result of instantiation may create a Controller class with ControllerType as a parameter. The feedback could be on/off, PID or, multivariate. Three classes could result: Controller (on/off), Controller (PID) or Controller (multivariate).

Objects hide the underlying implementation, so only the creator of the object knows the inner details of construction. All consumers must deal with an object through one its three distinct interfaces:

- The public interface. This is the interface that is visible to everyone.
- The inheritance interface. This is the interface that is accessible by direct specialization of the object.
- The parameter interface. In the case of parameterized classes, parameters must be supplied to create an instance of the parameterized class.

Aggregations

Aggregations are used to build composite objects. Composite objects are objects that contain parts which themselves are objects. The associations represent relationship between the objects. For example, a sensor may be composed of a number of components, including a transducer.

Specialization

Specialization is the process of creating a new object based on a more narrow definition of an existing object or an object that is directly related to, and more narrowly defined than, another object. For example, a temperature sensor is a more specialized type of a Sensor.

Abstract Classes

Abstract classes are created so that they can be inherited by others. While mostly no instances are created, the individual characteristics are made available to specialized classes via inheritance. For example, the temperature sensor class will inherit all attributes and methods specified in the sensor class, as well as any relationships that sensor class might have with other classes.

Properties of Objects

All objects have three important properties: inheritance, encapsulation, and polymorphism.

Inheritance

Inheritance means that an object inherits part of its behavior and data from another object. For example, a positive displacement pump (PD) object might inherit the characteristics of a pump object, such as equipment ID, line size, and discharge pressure, as well as methods such as StartPump and UpdatePumpStatus. Objects can also have noninherited characteristics. A PD pump could have attribute StrainerType and method FlowRate.

Encapsulation

Encapsulation means that both the data and the methods for an object are contained inside the object. Encapsulation provides modular structure to the software. In procedural programming methods, if the data output format is changed on a procedure and that data is being consumed by other procedures, the procedures may not function correctly. In object-based software, the only legal way to talk to an object is through messages. If the method needs to be changed, as long as the interface is preserved, the change will be invisible to other objects in the system.

When objects encapsulate data, misuse of the data by other objects can be prevented. This helps in making the software more stable and is the most tangible benefit of object-oriented software.

Polymorphism

Polymorphism means having multiple forms. Objects could have different methods for responding to the same messages. The sending object does not know the difference, but the receiving object interprets the message and provides an appropriate behavior. Polymorphism is the underlying principle that helps us with object-oriented user interfaces. For example, a DeviceDisplay object could send the message "calibrate" to any receiving object. The receiving object could be a PressureSensor object or a pHController object. The message "calibrate" will invoke the same method within the receptors, but the result will be different. The PressureSensor may just do a zero adjust, whereas the pHcontroller may launch a sophisticated two-point calibration method.

Object-Oriented Programming

Object-oriented programming is the application of object-oriented principles to computer programming. To be a true object-oriented programming language, encapsulation, inheritance, and polymorphism must be supported. Here we look at the some of the object-oriented programming languages. These languages may not be used to define control applications, but are being used here to develop process automation systems.

C++ and Smalltalk are the most popular object-oriented programming languages. C++ is the object-oriented version of C language. C++ uses compile time binding, also known as static binding. The specific class to which the object can belong must be specified by the programmer. This provides high run time efficiency and generates a minimum of run-time code.

Smalltalk, on the other hand, uses run-time binding, also called late-binding. When a message is invoked, the appropriate class library is searched for implementation of the operation fitting that message. With dynamic binding, nothing about the type of object needs to be known.

C++ has been standardized, but SmallTalk language implementations vary between vendors. The most popular commercial versions of SmallTalk are:

- VisualWorks
- SmallTalk/V
- VisualAge

VisualWorks was developed by ParcPlace, which grew out of the Xerox PARC project. VisualWorks is platform independent. SmallTalk/V is smaller and simpler, and currently supports Intel platforms only. VisualAge is IBM's version of SmallTalk optimized for IBM platforms.

Java is another object-oriented language that is becoming popular. The Java language is interpreted just like SmallTalk. The Java compiler generates "byte-code" instructions that can execute on any platform.

Programming languages are simply tools. The choice of language for OOP depends on development preferences. With the popularity of object technology, class libraries are becoming commercially available. Unlike libraries for conventional languages, classes can be modified by inheriting functionality from vendor-supplied classes, without modifying the vendor-supplied code.

Class libraries for GUI's and real-time communications are already available.

When selecting a programming tool, you should consider the following:

- Concurrency
- Persistence
- Garbage collection

Concurrency is important when you have multitasking application requirements. The objects seem to imply that concurrent operations are permissible, but implementation provides support only for sequential and synchronous computing. Smalltalk provides process and semaphore classes for multitasking.

Persistence allows objects that are no longer required or used in an application to be stored and retrieved later.

Garbage collection is the process by which temporary objects are discarded. Some objects live for nearly the full life of the application program, whereas others are temporary.

Object-Oriented Analysis and Design

The object-oriented paradigm for software engineering has begun to mature. Structured analysis, a model-building activity, helped to depict data and control flows and to logically partition the systems based on function and behavior. Although it worked very well for previous applications, the methods became strained with newer and more complex applications. Structured analysis is not very well suited to object-oriented programming. Object-oriented analysis methods are being developed and becoming accepted in analysis and design activities.

The primary need for object-oriented analysis methods arose because of the need to have object-oriented thinking throughout the development process. In object-oriented development, more time is spent on analysis and design activities as compared to structural analysis methods. The analysis process can promote reusability. Object-oriented methodologies are evolving, some attempting to provide full life-cycle development. Object-oriented analysis provides formal methods to analyze and design new complex systems. The previous methods of analysis provided us with reliable data processing applications. Object-oriented analysis methods are used to get a concept centric view as opposed to the function-centric view provided by structured analysis methods.

Although none have become a standard yet, several formal methodologies are available today. The most popular ones are:

- Booch
- Coad–Yourdon
- Fusion
- Jacobsen
- Martin–Odell
- Rumbaugh
- Shlaer–Mellor
- Unified modeling language (UML) by Booch–Rumbagh–Jacobsen
- Wirfs–Brock

All methodologies use a basic process to identify the subjects and objects and to define their behavior and relationships. The notations they use may be different.

Standards organizations are working to define object models. The Object Management Group (OMG), ANSI, and ISO are active in this area. OMG has published a document comparing different methodologies. ANSI X3H7 and ISO/IEC 107.46 are ongoing efforts to compare object models with available object methodologies. The unified modeling language (UML version 1.1) has become the standard for object-oriented development. It will serve both the analysis and design of software systems. The success of analysis tools will depend on support for code generation and CASE tools.

Object Database Management Systems (ODBMSs)

Relational database management systems (RDBMSs) help to separate the application from the design of shared databases. Data within an RDBMS is organized in two-dimensional tables. Complex data items are organized by joining rows from different tables and building artificial constructs. Structured query language (SQL) provides a means for expressing how data may be retrieved and manipulated.

Objects involve hierarchical arrangements of data. Storing and retrieving such data in a relational format requires decomposition and reassembly of the data.

ODBMSs do not require such a translation. Since the data is stored with its hierarchical relationships, it is easier to retrieve. The analysis and design datamodels do not have to be translated into a different database model. Object database management systems represent the next level in the technology of databases.

Components

Components are standalone objects that are often referred to as distributed objects. Distributed computing has resulted from the convergence of object and client–server technologies. This technology allows objects to be distributed over a network and components to interoperate. Components are typical "black boxes" without modifiable source code. The component does not support inheritance. It is a runtime executable that can be incorporated in an application. In distributed object systems, infrastructure must make it easier for components to self-manage and collaborate.

Object-enabling software infrastructure or mechanisms are required to make this happen. This means that a common interface needs to be defined and the connections between components need to be validated.

Objects interact with one another by passing messages to each other. During this interaction, the objects may assume the role of sender/receiver or, in other words, client/server. The infrastructure required is called an object request broker (ORB). The object management group (OMG) is a consortium that was formed in 1989 to define standards for distributed object systems. OMG's objective is to define an object management architecture (OMA). The OMA includes four sets of standards:

- Common Object Request Broker Architecture (CORBA)
- Common Object Services Specification (COSS)
- Common facilities
- Application objects

CORBA

CORBA 1.1 was defined in 1991. This specification covers the protocol a client application can use to communicate with an ORB, and a language for defining interfaces to objects (interface definition language). This specification did not specify an implementation. In 1994, CORBA 2.0 specified how ORB's from different vendors can interoperate. Some commercial implementations of the earlier version of CORBA specification are available.

ORB is the backbone of OMA and enables objects to make requests and receive responses in a distributed environment. A request (from a client) identifies an operation (method) to be performed on a server (target) object. Parameters, which may include references to other objects, are passed with their identifier. The ORB locates the server object and arranges for the method to be invoked on the object. The ORB is also responsible for any results that need to be returned to the client object. The ORB may use object services for processing requests/responses.

The client issues requests through client stubs or a dynamic invocation interface. Client stub represents an object operation, which a client invokes in a language-independent manner. Alternatively, a client can dynamically construct and invoke request objects that represent any object operation. On the implementation side, requests are received through skeletons. The skeleton provides an interface through which the method receives a request.

The interface definition language (IDL) is the language used to describe the interfaces that client objects call and implementation provides. An IDL is like an application programming interface (API) that specifies the set of operations with a set of input/output parameters for each operation.

The OMG has identified four examples of ORB's.

1. Client- and implementation-resident: the ORB is implemented as libraries (routines) resident in the clients and in the implementations.
2. Library resident (single-process): the ORB is implemented as libraries resident in the client.
3. Server-based: the ORB is implemented as a separate process, which brokers requests between clients and implementations.
4. System-based: the ORB is part of the operating system.

Common-Object Services

Common-object services extend the capability of the ORB in creating and managing objects in a distributed environment. Object services are packaged as objects by the ORB vendors. Some of the following services have been identified:

- Naming
- Event notification
- Persistence
- Transactions
- Externalization

- Relationships
- Life cycle

Common Facilities

OMG has categorized the collection of IDL components that define the rules of engagement for application objects. There are two facilities: horizontal and vertical. The horizontal facilities address four domains: user interface, information management, systems management, and task management. The vertical facilities address industry-specific domains.

Application Objects

Application objects can be thought of as traditional applications. Thus, one large object or a group of objects provide some set of functions for solving a problem within a specific domain, for example recipe management.

Distributed Computing Architecture (DCE)

The Open Software Foundation released initial specifications for distributed computing in 1990. The DCE was intended to provide facilities and services required to build applications on distributed computing environments. The suite of DCE services include:

- Remote procedure call (RPC) facility. The RPC facility provides the tools for building distributed applications using the client/server programming model.
- Threads facility. The threads facility supports creation, management, and synchronization of multiple threads of control within a single process.
- Distributed name/directory service. The distributed name/directory service provides a central repository for information about resources in the distributed system.
- Distributed time service. The distributed time service provides synchronized time on the computing nodes participating in a DCE.
- Distributed File service. The distributed file service allows users to access and share files without having to know the physical location of the file.
- Distributed security service. The distributed security service provides secure communications and controlled access to resources in the DCE.

- Diskless support service. The diskless support service allows a diskless host to acquire an operating system and configuration information over the network.

The DCE specification is fully defined. Some versions of the DCE are being used by large software vendors. Microsoft NT Operating system uses RPC as a foundation for COM+.

Distributed Component Object Model (COM+)

COM+ is a set of software objects that supports a set of functions to allow applications to connect to one another. The COM serves to connect a client and an object. After the connection is made the client and object communicate directly. COM+ defines a binary interoperability standard. COM+ also provides wire protocol for interaction among objects in a heterogeneous environment. This means that components written and packaged in a particular format per the binary and network standard can interoperate on the network. With COM, the applications interact with each other through function calls called interfaces. COM+ is controlled by a single vendor—Microsoft.

Framework

A framework is a reusable set or collection of classes that work together to provide a commonly needed functionality not provided by any of the individual classes inside the framework. A framework can also be thought of as a grouping of multiple objects to perform some function; they do not have to be reassembled in that same relationship every time they are needed. The ultimate goal will be to create frameworks that become business objects.

Summary

IRENE: "So object technology has some potential benefits in helping build software faster and making it easier to evolve. I understand that objects have three characteristics: inheritance, encapsulation, and polymorphism."

LARRY: "I really don't see how any of this applies to us, though."

IRENE: "Allen told us before that S88 was developed using object concepts. Do we need to understand object technology in order to apply these models?"

ALLEN: "The S88 committee uses object modeling to define data structures. It isn't necessary to understand object technology to use

S88 models. However if you understand object technology, you will be in a better position to analyze and define your requirements. It also helps to understand where technology is going so that you can minimize the impact of changes in the future. I could recommend several good sources on object technology "

IRENE: "And if we are going to apply object orientation, people like me need to be trained in it."

ALLEN: "Good point! Training is a very important part of all of this. Let's discuss it next."

Follow-through

Chapter

16

Retrain and Realign

STEVE: "We are now ready to commission the process. What have we done to assure that the operators will be able to operate the process?"

OWEN: "If it's anything like a typical project, nothing."

VICTOR: "This project is going to affect more people than just the operators. Are we ready to deal with all of the organizational impacts?"

IRENE: "We have a very good definition of how the process will operate. That can be used to train the operators. Based on what we said earlier, the operational spec will be an integral part of their job from now on. They need to be trained how to read it and what it means. The system we are installing is drastically different than anything that they are familiar with. It moves them into a role of managing the process rather than controlling the process. I think that based on that, I can develop some training for the operators."

VICTOR: "That's great, but what about the schedulers? We are totally changing their job. The system will now schedule batches into a batch management system. Those batches will then be executed automatically. We will have virtually eliminated the paper schedules."

ROBERT: "It's going to have a large impact on many departments. The QA department needs to know how to view the batch record, since it will now be electronic."

STEVE: "We are greatly increasing the load on the plant LAN. The IT department isn't used to supporting manufacturing critical applications and systems. We have introduced a much higher level of IT support needs. We need to increase the number of people in the plant IT support department or else we will get into trouble."

ALLEN: "You have recognized many of the ways in which automation will affect organizations and people. The following sections will examine this interaction in greater detail."

Individual Needs

Don't underestimate the effect of technology on an organization. When your child turns 16, do you put his or her bicycle in the garage, hand over the keys to your car, and say "you can drive now?" That's exactly what you are doing if you automate a process and don't teach everyone involved how it will affect him or her. Unfortunately, operators will usually find a way to operate their fancy new car (process) using their feet, Fred Flintstone style (Figure 16.1).

Although automation will not reduce the number of people needed to operate a plant, it will change the jobs and skills required *and that takes training!* By adapting to the different needs of automation, you can significantly increase the efficiency, reliability, and productivity of the process. By not adapting, you could increase the number of people required to run the plant, and decrease its efficiency.

An often-overlooked aspect of technology is its impact on an organization. If implemented properly, technology will change how your organization operates. The problem is, if you don't recognize and plan for it, it can be disastrous.

Using the strategic automation methodology, the operators are involved in the definition process. Just think of the benefits. They are learning how the process will operate automatically while they are helping to define how it would operate automatically. Talk about win–win! Additionally, they will now have a document written in un-

Figure 16.1 The Flintstones.

derstandable language describing the operation. This is all fine and good, but it isn't enough.

Organizational Needs

Here are some issues that need to be considered. You need to take a long hard look at how an automation project will affect your organization. Much infrastructure is necessary to support an automated process. This can be provided either through internal resources or through contract support. There are advantages and disadvantages to either approach. This decision needs to be made based on the needs of the plant, the available local support resources, and the company strategy.

Do your operators have computer skills? Do they understand instrumentation? If a valve sticks will they be able to isolate the problem? Do they understand what information is available to them through information systems and historians? Are they now trained to improve the process operation rather than operate the process? You have now given them a totally new and more rewarding job, but you need to give them the training and tools to perform it.

The changes go far beyond the operators, though. How will the change affect maintenance? The process is now much more high-tech. The shops now need to understand all of the interrelationships within the process before working on a piece of equipment.

As standard off-the-shelf computers are beginning to be applied to automation, your computer systems department needs to understand how the machine affects the process. They need to understand how the local area network can affect manufacturing. Do you have the level of support in place to provide for the reliability necessary for a manufacturing process?

Standard operating procedures and product recipes are now on-line dynamic documents. How does this affect the people who were developing them? Do you need as many people as before?

The scheduling of production is now done dynamically in conjunction with the automation system. Do you need the same number of schedulers? How do the schedulers need to be trained?

Plant management can now view manufacturing data in real time. How do they need to be trained to use this information? Do you still need the same number, or type of managers? Can this real-time data be used to empower production employees to make more of the decisions?

You must manage the organizational change that goes along with the automation. Unless the change in technology is minimal, there will be training and organizational change required. As we said earlier, an automation project will typically not reduce the number of people required within a plant, but it will change the skills required. If you don't manage this change, you could end up with more people than you need, because you haven't updated the skills of your people.

Documentation

We have defined many forms of documentation. One of the most important documents for the operators is the operational specification (Op Spec). This shows them how the automation will be operating the process. They need to know how to read and understand these Op Specs.

In addition to the Op Specs are the Operational Manuals. These tell the operator how to interface with the automation. They need to be well versed in how to operate and troubleshoot the system.

Process or equipment SOPs quite often include information about the process control. Since our Op Spec is now a controlled document, and the operators are fully trained in how to read and use it, there is no need to duplicate this information. Equipment SOPs should simply reference the appropriate Op Specs.

A last class of document that is sometimes used is the product SOP. This tells the operator how to use the process (both equipment and automation) to make a specific product. It is part of the product master recipe. Product SOPs should only reference the appropriate Op Specs and equipment SOPs.

Summary

OWEN: "So what you are saying is that I'm going to lose my job to someone with higher skills?"

ALLEN: "That's not what I said. The skills to do your job are definitely going to change, though. In today's economy, everyone needs to continually update their skills to remain competitive. There will be more opportunity for career advancement and more challenging jobs."

OWEN: "What if I can't learn the new systems?"

IRENE: "Owen, you won't have any problem. If the system that we have designed isn't intuitive, then we haven't done our job. Besides, you were involved in the design. You didn't let us design a system that you couldn't operate, did you?"

MARK: "It seems like much of what is needed now is troubleshooting skills."

STEVE: "That is true. We have designed a system that will operate during all circumstances that we could anticipate. The operators will now take care of the rest of the circumstances.

Allen, on a separate subject, should the people supporting the automation be part of manufacturing, or in a separate IT function?"

ALLEN: "That totally depends on the company and plant. No matter what, they should be measured against the same measures as is manufacturing. Although they are a support function, everyone in the plant is responsible for making product. The automation is a critical part of making product. The measures for the automation support personnel should be such that it ensures that making product is the goal and the control system is a tool to achieve the goal."

STEVE: "Based on what you said early on, I'd assume that there is a difference between the people supporting the automation and the people supporting the control system."

ALLEN: "Excellent point! The people supporting the automation should have the same goals as manufacturing. Quite often, the person supporting the automation will be the manufacturing engineer. The control system is another story. The goal of the system is to maintain reliability and on-line time. The goal of the automation is to make product."

ROBERT: "What about simulation? Should it be used for training purposes?"

ALLEN: "Absolutely, if it is appropriate. If the complexity of the process is such that the operators need training to be able to react to situations before they have actually encountered them, then simulation should be used to train the operators. This will then prepare the operators respond to upsets."

17

Building Recipes

PATTY: "Now that we're getting ready to commission the process, shouldn't we be developing our product recipes?"

STEVE: "That's true. Everything that we have done up to this point has been based on the physical process. Are we going to have the product recipes ready in time for start-up?"

IRENE: "Based on how we built the automation, developing product recipes shouldn't be difficult. We evaluated our equipment procedures against the products that we are going to manufacture with this process. Now all we need to do is build a recipe procedure for each product."

ALLEN: "I have a suggestion. In order to make reusable modular recipe procedural elements, it is better to start by building a foundation of operations first."

Building Recipe Procedures

When we discussed modularizing the physical process, we said that it was best to work from the top down. First, define the process cell, next the units, next all equipment modules, and lastly all control modules. This helped to enforce a modular design. The opposite approach is better when dealing with procedures.

By using a bottom-up approach to designing procedural elements, it is much easier to end up with reusable, modular procedures. You should have built your equipment phases while you were defining the physical process and equipment procedures. Now you need to map those equipment phases to recipe phases. The recipe phases are defined in batch management software. They are the foundation on which you will build your recipe procedures.

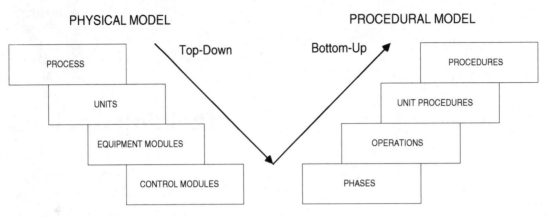

Figure 17.1 Methodology steps.

Once you have defined your recipe phases, you should begin looking at product needs. Operations are the highest procedural level that can typically be built with any product independence. Some typical examples of product-independent operations are purge, react, and transfer. Each of these could be built in such a way that they would work for any product.

Once you have built your product-independent operations, you next need to define the remaining operations. These will typically contain processing information (e.g., how raw materials will be loaded) for making each specific product.

Next, unit procedures that can be used to coordinate the activities of units within the process cell are built. Rarely will a procedure for a given product contain only one unit procedure for each unit. By subdividing the unit procedures, it becomes easier to manage the allocation of resources.

Imagine the following scenario. There is a process with a network of five units (see Figure 17.2). There are two premixers, two mixers, and a filter. Each premixer can feed either mixer, and each mixer can feed the filter.

If you build a recipe procedure that has one unit procedure per unit, the units will by default be allocated to the batch as soon as the batch begins execution. If there is still a batch in the filter, another batch cannot be started in either premixer. Most batch engines have mechanisms that can minimize this impact, but it is helpful to build unit procedures that don't acquire the equipment until it is needed (see Figure 17.3).

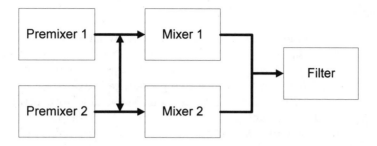

Figure 17.2 Sample recipe process block diagram

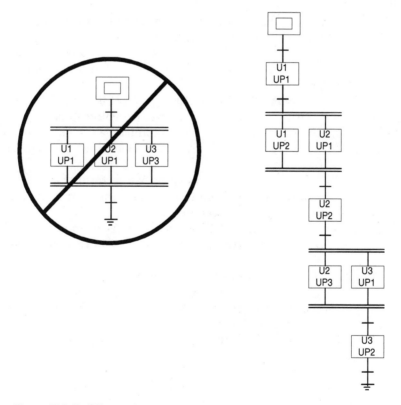

Figure 17.3 Building recipes.

Class-Based Recipes

A recipe is used to make a product. Whenever possible, master recipes should be defined without any references to what specific equipment entities it will run on. When a control recipe is generated, equipment is allocated, and the recipe procedure is executed.

A recipe procedure does not know how to control the equipment. Control is not a part of the recipe. The recipe procedure is linked to equipment procedures at some level to perform the actual control. This separation of the equipment control from recipe provides flexibility in system implementation. The control recipe procedure can be detailed with references to recipe phases (heat, cool) or to high-level recipe procedure (make batch). Depending on the equipment control, and control recipe, the linking can be done at the phase, operation, unit procedure, or procedure level.

Example

PATTY: "I'm somewhat confused as to how we are going to design procedures that are reusable."

ALLEN: "Well, let's talk about the process. Is there any work that will need to be done in the reactor that is the same across all or some products?"

DONNA: "Not really. Each product is unique."

OWEN: "You mean that we are going to start transferring product differently, based on what product we are going to make?"

PATTY: "Well, no. But what does that have to do with designing recipes?"

IRENE: "I see Owen's point. We certainly be able to develop unit procedures that will coordinate the transfer, independently of what product is being manufactured."

ALLEN: "Let's talk about a simpler example first. The reactors have a pressure control using nitrogen. Will they be purged?"

PATTY: "Yes."

ALLEN: "How will the purge work?"

PATTY: "We will measure the oxygen content of the reactor. If it is above the setpoint, we will pressurize the Reactor to 10 psi and then vent to 2 psi" (Figure 17.4).

IRENE: "Why couldn't we make that an operation, and use it for all products?"

DONNA: "OK, so there are some reusable operations."

STEVE: "The CIP system should also work the same for all products. How will that work with the individual units, though?"

ALLEN: "We need a clean phase for the specific unit in parallel with

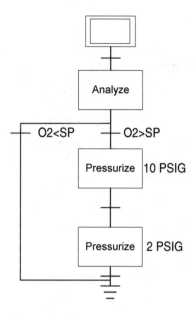

Figure 17.4 Purge operation.

the CIP phase. Since the CIP system is an equipment module, the CIP phase is acquired by a unit and executed in an operation by that unit."

PATTY: "So it would look like this?" (Figure 17.5).

ALLEN: "Looks good."

DONNA: "Can we build a full product procedure now?"

IRENE: "I'd like to. Do you have a recipe, Donna?"

DONNA: "Yes. Here's a simple one:

       Mix into Reactor:
          100 lbs of Raw Material 1
          150 lbs of Raw Material 2
          300 lbs of Raw Material 3
       React for 20 minutes by:
          Heating to 150 °F
          Continuing to mix
          Adding Catalyst
       At same time, prepare the distillation dolumn.
       When done, transfer the contents through the distillation column at 180 °F.
       Send the overheads to Tank 1 and the bottoms to Tank 2."

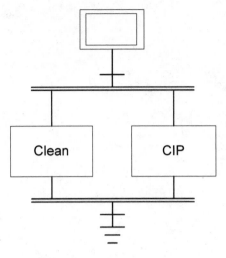

Figure 17.5 Clean operation.

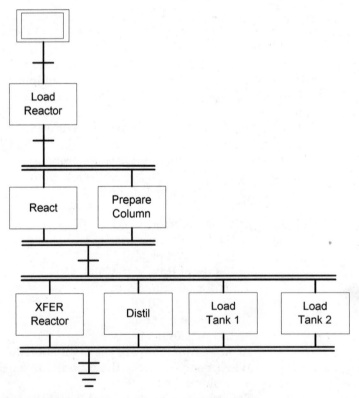

Figure 17.6 Example procedure (see Appendix C).

IRENE: "Would this work as the procedure?" (Figure 17.6).
DONNA: "That looks like it should do it."

Summary

STEVE: "It's good that we have started building recipes because we are ready to commission the process."

ROBERT: "Of course, we need to validate it first."

ALLEN: "Since we are ready to validate the process, let's discuss validation next."

Chapter

18

Validation

LARRY: "Validation is on the critical path for our project. Why do we have to wait until after installation? We seem to spend a lot of time doing validation. Is our approach correct?"

VICTOR: "I've heard that we have a very good validation program in place. Since we will be making products that are regulated, we have to make sure the system is qualified before we can start shipping products."

ROBERT: "Yes! We do have a validation program in place. We followed the lead of industry organizations attempting to interpret the regulations."

LARRY: "It seems like the amount of testing and validation we have done in the past is enormous."

ALLEN: "You're not alone. The regulations aren't clear even for some individuals who work very closely with them. The requirements are continuously changing. You have to realize that a validation effort depends on several factors, including the complexity of the equipment and the function it performs. Unfortunately, there isn't a one size fits all program.

The strategic automation methodology has qualification built in. The methodology takes a life-cycle approach to automation. The validation activities are an integral part of the methodology. When validation is approached in this manner, it's no longer an ad-hoc event in the life cycle of the process, but becomes a part of the process itself."

ROBERT: "In your opinion, what makes a good validation program"?

ALLEN: "I'm just going to discuss the validation of automation systems. An automation system consists of a computer system and the equipment that it controls. We know that a computer system is an assembly of hardware and associated software collectively designed to perform specific functions. A good validation program takes the environment in which the computerized system will operate into account and should address all aspects of the life cycle of the system, including continuous improvement and decommissioning. Let's review what we said earlier about validation (Figure 18.1). Qualification is the process of establishing documented evidence that the system operates as specified."

LARRY: "That's all fine. But how do we know what level of validation is required?"

ROBERT: "We classify the system based on two criteria: regulatory needs and complexity of the system."

VICTOR: "I thought we had to validate all systems."

ROBERT: "When we do the classification, we look at the following evaluation criteria:

- Does the system affect the quality of the product?
- Does the system impact safety or the environment?

Based on this evaluation, we recommend a validation approach."

ALLEN: "Does this mean that you recommend validation for some systems but not others? Does that mean that you have to make judgment calls?"

ROBERT: "Yes, there have been instances where we've made judgment calls. We know that validation is a risk versus cost balancing act. You have to realize that penalties and significant costs could result from a recall of a drug or its active component."

VICTOR: "Can you give us an example of a system on which you made a judgment call?"

ROBERT: "We had a labeling system that printed product data on bulk drums. The drums were shipped to a warehouse and then on to customers. The product data was coming from a validated system. A

Rule of Thumb for Validation

1. Say what you are going to do.
2. Do what you said you were going to do.
3. Manage all changes.

Produce independently verifiable evidence .

Figure 18.1 Validation principles.

label verification procedure was in place. After careful analysis, we decided it did not make sense to validate this labeling system. However, we are ultimately responsible for assuring systems used in manufacturing are in compliance with regulations."

VICTOR: "During this project, we've followed the strategic automation methodology. How will this work with our existing validation program?"

ALLEN: "I need to know more about your approach before I make any recommendations. Robert, can you tell us more about your existing validation program?"

ROBERT: "The validation program applies to the complete process life cycle. Our process life cycle spans discovery, development, production, improvement, and decommissioning. The computer system validation is part of the overall validation effort and uses a life-cycle approach to validation. Before I go into details, it may be worthwhile to review some validation concepts."

Computer-Related System

A computer-related system executes a "process." Here process is not restricted to manufacturing but could include laboratory processes, documentation, or data management. The computer system is composed of hardware and software controlling the operation of the computer. The controlled function may be composed of equipment to be controlled and operating procedures that define the function of such equipment, or it may be an operation that does not require equipment. We only consider equipment that performs a GMP function. The computerized system functions in an operating environment that includes information input from operators or other sources, input of the material to be processed, the physical environment, plant support utilities, and any other factor that might affect the performance of the computerized system.

A *computerized system* is composed of the *computer system* and the *controlled function.* A computer-related system is a computerized system performing its intended function in its operating environment (Figure 18.2).

There are three widely accepted methods of validation. The first, *prospective validation,* involves developing a preplanned protocol to establish documented evidence. The second, *concurrent validation,* establishes documented evidence during the actual implementation of the process. Last, *retrospective validation* establishes documented evidence based on the review and analysis of historical information. The retrospective validation approach has been successfully used on some preexisting computerized systems that predate current validation

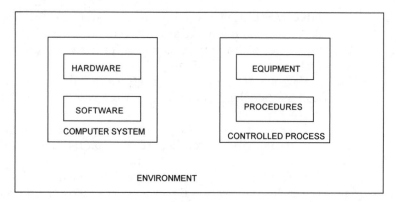

Figure 18.2 Computer-related system.

guidelines, but prospective validation methods are more typically used.

Validation of computerized systems became a legal requirement in 1978, when the cGMP regulations were published. The interpretation of these regulations has varied widely. The Pharmaceutical Research and Manufacturers of America (PhRMA), formerly the PMA, with its Computer System Validation Committee (CSVC) has played a major role in clarifying regulations. In 1986, CSVC published its "life-cycle approach to validation of computer systems."

Validation Life Cycle

PhRMA also provided guidance on validating preexisting systems using a modified life cycle and retrospective approach (Figure 18.3).We extended the life-cycle model and have come up with a five-stage model for validating computer systems. The five stages are:

- Plan
- Define
- Design
- Qualify
- Evaluate and support

Planning Stage

In the planning stage, a master validation plan is created. This validation plan is the primary tool used to align and integrate the valida-

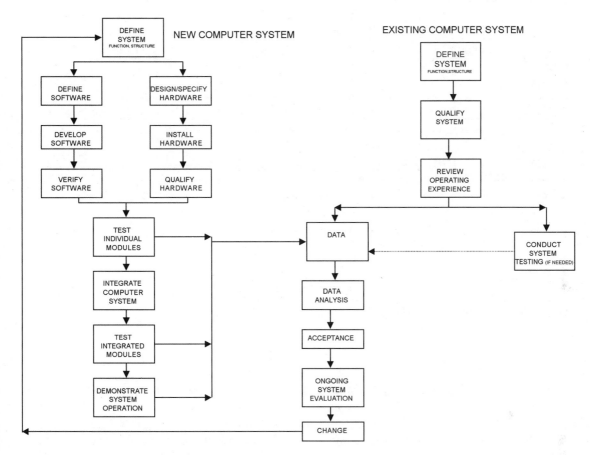

Figure 18.3 PhRMA's life-cycle approach.

tion program with the design/engineering and construction activities. A schematic of the validation plan (Figure 18.4) shows components and activities required to develop the validation plan.

LARRY: "Last time, we generated a validation plan for the whole facility. For a project manager, like myself, it exposed me to the detailed intricacies of the validation program. It also helped us to share information, resulting in better coordination. However, we did experience some scheduling problems in the execution of the plan during start-up due to the extensive reprogramming. The protocols had to be rewritten."

VICTOR: "Can you tell us how you organized the validation plan?"

LARRY: "The validation plan that we put together had a lot of infor-

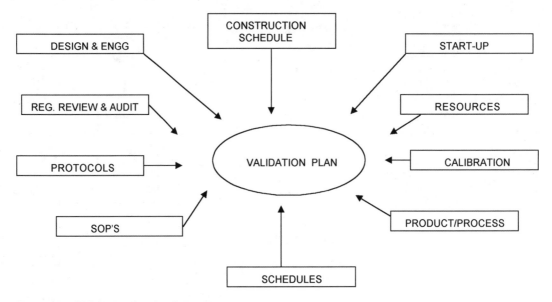

Figure 18.4 Validation plan development.

mation. It defined the roles and people responsible for different parts of the program. It identified equipment, systems, and processes that were to be validated. Patty and Robert helped me identify all of the regulatory needs and the level of validation required. We then identified the acceptance criteria for each system and process to be validated. Each individual was asked to define the acceptance criteria for his or her system. We then listed all of the validation SOPs and the method to be used to execute each validation protocol. Last, we developed a tentative project schedule."

VICTOR: "How did you manage all of that information?"

LARRY: "We drew Gantt charts for each part of the validation plan. The project management software I used was for planning purposes only. For the next project I would integrate the validation plan in my master project schedule. This will make it easier to calculate project schedules based on commonly accepted methods. I am responsible for the cost of the project as well, and everyone knows that delays cost money."

ROBERT: "I think the problems we ran into during qualification can be attributed to several factors. For one, the people responsible for the actual validation of the system could have done a better job."

Definition Stage

The first stage also includes the functional specification of the system. In the Define phase, the User Requirements Document (URD) is assembled. From this document, the design requirements and functional requirements must be exposed. A good URD document will contain a written description of the process. There are several advantages in having the system defined up front. The focus of system definition is to outline "what" a system will do rather than "how" it will do it. It is interesting to note that regulatory agencies often cite lack of definition of computer systems as a problem.

Design Stage

The design will form the basis for the validation of the system. The set of design documents will help to identify all measurable and determinant factors that may affect the performance of the computerized system and its critical parameters.

The design documents will also list how the automatic verification criteria of regulations will be met. The design document will make references to various other documents, such as the URD, P&IDs, and SOPs for detailed information. The complexity of the system might determine how the design documents are structured. Hardware design and software design documents will be required if considerable integration and application software are to be developed.

During the design phase, the test plans are developed. A test plan outlines how the validation criteria will be met. For complex systems, various components of the system could be integrated after installation at the plant or come prepackaged from the vendor. The test plan may be executed at the integrator's facility or at the plant. Software tests are an important part of validation of computerized systems. Software testing may involve structural and functional testing. Structural testing is the detailed examination of the internal structure of the code. This testing is accomplished through reviews or walk-through with the development team. The main purpose of structural testing is to ensure the integrity of the software. Usually, structural testing involves:

- Review of code to determine whether the code actually adheres to established standards
- Review of code documentation
- Analysis of all logic paths
- Inspection for dead code

Functional testing of software is a thorough and systematic comparison of observed output values of a program with expected output values for a specified range of input values or parameters. Often, functional testing plans are derived from the system specification, in which descriptions of functions are available in sufficient detail.

If the user develops software, it is imperative that quality assurance (QA) activities are undertaken in order to provide adequate confidence that specifications are met. Usually the QA ensures that software quality is managed throughout the software life cycle. QA includes defined processes, standards and practices, and procedures and guidelines for software development, testing, and maintenance.

The qualification protocols are written at this stage. Protocols have to be approved before they are executed. Protocols for installation qualification (IQ) and operational qualification (OQ) are established. Each qualification may break down the tests into components and assign test scripts for each component. The test script includes the purpose of the test, instructions to execute the test, the test procedure, expected result, and/or the acceptance criteria. The test script should account for normal and abnormal conditions.

Qualification Stage

During the qualification stage, other activities such as installation, integration and prestart-up occur as the validation protocols are executed. The results of the tests are recorded in the test script.

The installation qualification is the documented verification that all important aspects of installing hardware and software have been verified. The installation qualification also verifies that appropriate user manuals, drawings, calibration reports, and SOPs are available. The hardware portion typically tests the following items:

- Environmental conditions (power, grounding, temperature, humidity)
- Electromagnetic interference
- Wiring schematics and terminations
- Instrument database
- Preventive maintenance and calibration programs
- Network connections

The software portion includes the following tests:

- Correct installation of application software

- Document and verify version of applications along with name, size, date, time, and attributes
- Back up copy of installed software

The operational qualification process is the documented verification that the system operates in accordance with the specification throughout all anticipated ranges. Operational qualification may be performed on the integrated system or on each subsystem and includes the identification of all important parameters, their anticipated ranges, and appropriate acceptance criteria. Tests are also to be designed to verify and challenge the following:

- Safety interlock systems
- Timers
- Alarms
- Security
- Redundancy
- Network communications
- Safe-state commands
- Operator data entry

When the qualification of the computer system is complete, the process can be validated. Performance qualification activities start as the summary report and certification "ready for production" is made. Usually, all the deviations noted during the execution of the qualifications are resolved. The qualification report must include:

- Test protocol results
- Summary of the tests indicating confidence and suitability of the system
- List of deviations
- Criticality of the deviations

After the summary report is complete and the deviations have been resolved, the system is certified "ready for production." After the certification is complete the performance qualification (PQ) starts. PQ is the documented evidence that a system performs its intended function, as per the process specification and the system specification, while operating in its normal operating environment. The PQ is always done on the integrated computerized system.

Evaluate and Support

This phase involves the maintenance of the computerized system over its life cycle. The maintenance activities may include both planned and unplanned changes. The change control activities are usually controlled by a procedure. A well-established change control program is required to maintain the validated state of the system. Change control procedures describe steps to authorize, perform, test, implement, and document changes.

Maintenance and calibration procedures should be written and approved for every system component of a computerized system, in order to maintain the integrity and accuracy of sensors and instruments.

Preventive maintenance (PM) schedules should be followed. A good PM program will have detailed instructions for frequency of inspection, cleaning, and repair of equipment and devices. Other procedures for computer system security and disaster recovery plans should be in place. An equipment log is required for tracking any maintenance-related activity.

ALLEN: "The validation program seems to be well thought out. I would like to map this validation program to our strategic automation methodology. I would also like to make recommendations to improve the program."

The strategic automation methodology outlined in Chapter 2 accommodates all elements of a validation program. During the project specification phase, the objectives for the process are set. The validation plan should be defined at this stage.

The operation definitions are accomplished during phase II. The phase deliverables include operation specification and system specification. Operation specification is the URD. Detailed validation plans can be created after the system specifications are available. The operation specification also forms the basis document for performance qualification. Phase III deliverables include the design documents. The design documents form the basis for the validation protocols for installation and operational qualification. Phase IV is the implementation and qualification. In this phase, installation, start-up, and qualification activities occur. The continuous-improvement phase manages all changes. The documentation you generate for the process automation system is all you need for carrying out a successful validation program.

ROBERT: "I think this program is very good. What additions do you foresee?"

ALLEN: "For a typical project, how do you assign validation responsibilities? Is there a clear understanding amongst the team about validation deliverables and why it needs to be done?"

ROBERT: "There are clear accountabilities for validation. When we put a new facility together, we get involved at the earliest stages of the project. The project manager is responsible to get all the deliverables for the validation program. We have trained our professionals on approaches to validation."

ALLEN: "Do all professionals know about your validation SOPs?"

ROBERT: "We make good efforts to inform the professionals on the validation strategy and policy. The validation SOPs define and provide detailed instructions for executing validation tasks. There are SOPs for developing qualification protocols, their execution, and the levels of approval and review of validation documents. The validation policy is updated based on changes in the regulatory environment and communicated regularly to the professionals."

ALLEN: "It is good to assign clear responsibilities for validation, in either a project setting or in the maintenance phase of the project. The responsibilities must be clearly communicated to prevent omission or overlap of efforts. For example, the process automation system may interface with electrical devices like motors. When qualifying the device installation and operation, it helps to coordinate this action to avoid redundancy. If not, clear delineation of systems should be made. You can represent pictorially the different SOPs that need to be followed. This can help clarify the interrelationships that are involved in the process automation system and be made a part of the validation strategy.

In this sketch (Figure 18.5), I have used a scaleable process automation System that could be in any facility. The logical boundaries of the systems are shown. Is there uniform policy on acceptance criteria for validating process automation systems?"

ROBERT: "We define the acceptance criteria for each system. The first major decisions are on the level of validation. For this, we work very closely with the concept engineering team."

ALLEN: "How do you decide on criteria for vendor audit?"

ROBERT: "When parts of the computerized system are provided by vendors, we independently audit them before the vendor is selected. We have an SOP on how to perform a vendor audit."

ALLEN: "If system strategy and architecture are in place, the process of vendor audit can be tied to the selection of the vendor and the integrators. The GAMP draft guidelines provide a good checklist for different types of audit, including technical ability, validation status, and quality. If a sound strategy for automation is adopted, you can avoid duplication of vendor audits."

VICTOR: "I see your point. During our last project, the concept engi-

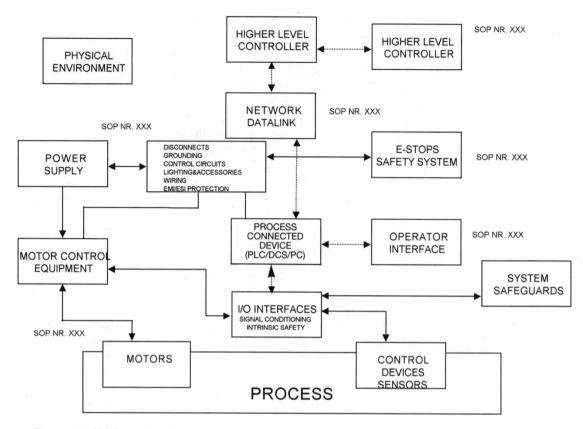

Figure 18.5 Validation boundaries.

neering team spent some of their valuable time traveling to perform audits."

ALLEN: "When the validation plans are generated, does everyone know exactly what needs to be done?"

LARRY: "Are you referring to installation or operational qualification? My team members say that the validation SOP was exhaustive and asked for tests on things that didn't really apply. During start-up, the internal audit team recommended more qualification tests than were listed in the validation SOP. We had to hire consultants to get them done."

ALLEN: "This is a problem with many people working on the program. To avoid surprises such as this, it is good to develop a qualification plan. There should be separate qualification plans for IQ and OQ. These plans can be in the form of a checklist. By selecting items on the checklist, you will also get a good idea of effort involved. The plan helps to quickly identify which aspects of the system need to be qualified.

ALLEN: "What happens to the documentation after the validation is complete?"

ROBERT: "The documentation is assembled and sent to the business owner of the process. All documents generated during the support stage are sent directly to the business owner."

ALLEN: "Do you have a consistent manner in which documents are assembled?"

ROBERT: "We do recommend a numbering scheme in our validation SOP. We do not track the documentation for compliance with those procedures."

ALLEN: "It always helps to have a consistent method to uniquely identify the documents during the entire life cycle of the process automation system. This will help to manage the system better. Documentation is not just for the express purpose of meeting some regulatory need. You can get a wealth of data about your systems from them. However, this benefit can only be realized if the documents are diligently assembled. Change management can also help in better definition of future systems."

ROBERT: "How do you determine how the document needs to be organized? We get several calls from our business owners. Each one seems to invent a unique way to organize their portfolio."

ALLEN: "The easiest way to organize documentation is by a unique system. An ID number should be assigned for each computer system and the associated functions that it enables. For example, if a change in application software is made, the associated change documentation should be appended to the validation portfolio. All related documents must be updated. The strategic automation methodology can help enforce change management. This way of organizing documents can have drawbacks, too. Other regulations may require a different kind of document organization. To accommodate the changing needs, it helps to cross-index the documents as well. Each facility will have unique needs for documentation. As long as this is recognized, the documentation itself will add value to the process."

Summary

STEVE: "Even though the validation is a lot of work, I think it is worth it. During my last start-up, the process started up so smoothly."

IRENE: "Well, it should have. We spent a lot of time defining how it should run, and we made sure, through the validation, that it did exactly what it should."

Reaping the Rewards

19

Continuous Improvement

IRENE: "I have a question. How are we supposed to assure that the project will be successful if we are pulled off onto another project before it is completed?"

STEVE: "I agree, not only is the design team disbanded too soon, but the start-up team is usually disbanded before the automation is fully operational. Victor, can we get commitment from you that we will be able to keep project teams together until the objectives have been met?"

VICTOR: "Absolutely! We will set up a feedback structure so that we can assure that the project results are validated before the project team is disbanded. Why are project teams disbanded before the project is completed, anyway?"

LARRY: "Usually, it's because the project is either out of budget or out of time. Sometimes some of the people are needed on other projects."

ALLEN: "Do you think that this new methodology will help to keep a project on track so that it doesn't go over budget or schedule?"

LARRY: "I certainly hope so! I guess only time will tell, though."

IRENE: "I, for one, believe that this will help to assure that the automation works properly one way or another. The reduction in rework should help to keep the project on budget and schedule."

Project Completion

An automation project is not complete once it is installed. As the axiom for validation goes "say what you're going to do, and do what you said you were going to do." This concept has two parts. The first is to assure that that the specification is correct. The second is to assure that the system performs as specified. These can be done before installation through simulation or with a live process.

An automation project requires long-term commitment to make it work. The first attempt at automation is unlikely to be perfect. All too often, a good automation strategy is scrapped due to early failures. The project cannot be considered complete until the original objectives have been met. That means that the members of the team must not be pulled off onto another project until the objectives have been met.

The only way to know for sure that a project is complete is to measure it against the project objectives. We can easily tell when this occurs, since we have developed an automation objective for the project that was measurable, specific, and achievable. The only alternative to meeting the original automation objective is a change in the automation objective. This can be due to a change in scope of the project, or a realization that the automation objective was unrealistic. Whatever the case, the automation objective needs to be updated to reflect the change. When the project is complete it is still measured against that objective.

A team implementing an automation strategy should be recognized and rewarded based on the achievement of the automation objective within the schedule and budget. All too many times, the team is measured and rewarded based on the schedule and budget alone. This can lead to projects that are completed on schedule and within budget, but do not achieve the desired results. A project that doesn't meet the original automation objectives may no longer be justified. Therefore, the fact that the project was on schedule and within budget is irrelevant. This is not to say that schedule and budget are irrelevant, just that the objectives must be reached before the schedule and budget are meaningful.

Another consideration for the completion of a project is the comfort level of the operators. Not only do they need to be trained on the process and the automation, but they need to be comfortable with how it works. This will take time and may require some changes to the designed system. If the operators are not comfortable with the automation, they will find other ways to do their job. This will sometimes mean that they override the automation. Since the project was justified upon the premise that the automation would add value, this is the same as the automation not performing properly.

Although most nonpharmaceutical processes do not need to be "validated" in the strict sense of the word, there are still some strong reasons to validate your project and automation. Simply stated, validation is a process through which you prove that your actual results match or exceed your desired results (see Figure 19.1). This is good

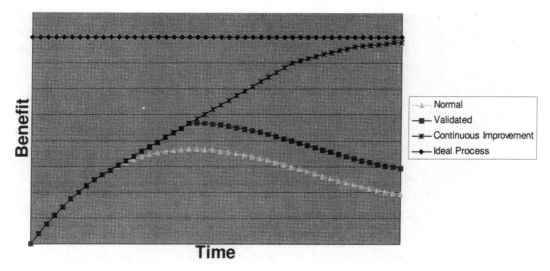

Figure 19.1 Benefit over time.

engineering practice. We typically do much of this through a checkout during start-up. The pharmaceutical industry uses a structured validation methodology to be applied to these checkouts. By imposing this structure, we define up-front what will be checked, how it will be checked, and what the acceptance criteria are. After the process is fully commissioned, there will be changes required to maintain and improve its operation. Even a validated process is subject to entropy. Everything decays over time.

The bottom line on automation is that it isn't complete until the operators are satisfied and confident that it will work. This doesn't mean simply being operational. It must help them do their job better in some way.

Previously, when we discussed the methodology of applying automation, we discussed the support requirements. This is where those support requirements come into play. There are a two of ways in which this support can be achieved. You can add in-house expertise, or you can enter into a support contract agreement.

Management of Change

Over time, entropy and continuous improvement will require changes to the system. Good engineering practices, as well as regulations, necessitate a management of change procedure. This should assure that the appropriate people review the change before it is implemented. It should be documented to show what changed and how the system will operate after the change is implemented. The change should be tested to assure that it works properly, and has not adversely affected other aspects of the system. The procedure should also have provisions to assure closure on the changes.

Make sure that all appropriate people are consulted before a change is made. An operator should always review any change that is going to be made before it is implemented. Some of the people that may need to review the change are:

- Manufacturing engineer
- Operator
- Safety personnel
- Environmental personnel
- Quality assurance personnel
- Product development personnel
- Automation engineer
- Process engineering

There are also some questions that should be asked before a change is implemented:

- Could the change have an impact on product quality?
- Could the change have an impact on the safety of the process?
- Could the change have an impact on waste generation or release?
- Could the change have an impact on plant utilities?
- Will the change have an impact on the operating procedures?
- Will the change have an impact on the process conditions?
- Will the change need to be validated?

If any of these answers are yes or maybe, then the change probably should be implemented using a change management procedure. The change management procedure should assure that the change is:

1. Reviewed for its impact on the process (including upstream and downstream processes)
2. Approved by appropriate persons
3. Documented
4. Communicated to the operators and anyone else affected
5. Tested (or validated)

It is a good idea to have checklists for changes. These checklists are for things like what needs to be completed before the change can be implemented and what documentation needs to be updated. Anytime that something is removed or decommissioned, all documentation should be updated or marked obsolete.

The operational specification that was created during the automation design should be used as ongoing documentation. This should be updated any time a change is made, and used to train new operators on the process.

Summary

VICTOR: "Well, this was one of the most successful automation projects that we've ever had."

STEVE: "It certainly was. I have to admit that I was skeptical at first. I've seen a lot of automation fail in the past."

OWEN: "*You* were skeptical. *I* thought this whole thing was a waste of time. Now I think that it should be done for all of our processes."

VICTOR: "Not so fast, Owen. Remember, we have to have a business driver to justify the expense."

LARRY: "I can't believe that with all of the time we spent writing documents in the beginning, we didn't go over budget and we weren't late."

IRENE: "Remember, Larry, we usually end up spending much of our engineering effort on automation either rewriting it or trying to find ways to make it work. This time, it just worked and did exactly what we wanted it to do. It was the easiest project that I've ever been on. Well, OK, so I've only been on one other project, but this one was still much easier."

ROBERT: "Having change management built into the project certainly also improved the documentation. The operators were trained, and all of the documentation matched reality. Now we just have to have the discipline to continue managing changes."

DONNA: "What I can't believe is that we were able to implement that new product in less than a day. It was a product that I hadn't anticipated, and it used the process in ways that I never dreamed that we would use It."

PATTY: "I knew that we had designed the process for products such as that, but it was great to actually see it work without having to change the automation."

20

Succeeding at Automation

ALLEN: "Now that we've finished the project, let's review our orignial issues (Figure 20.1) and make sure that we have touched on everything."

STEVE: "We certainly talked about documentation and training."

PATTY: "And I don't think that any of us will ever forget about the automation objective."

- Documentation and training need to be defined up front
- Need clear understanding of automation objective
- The right people need to be involved in the design
- Automation needs to improve on manual process
- Follow industry standards and trends
- Need better way to define the operation of a dynamic process
- Need better way to define and deal with interlocks
- Need to manage alarms so that they are meaningful

Figure 20.1 Issue board.

IRENE: "I think that we talked about all of the issues in some form."
ALLEN: "Well then, let's review some of our key topics."

Automation Principles

The entire basis of strategic automation is to get our automation to meet our ideal expectations. Process control has tended to add complexity and limit the flexibility of our processes. Strategic automation helps to reduce process complexity through the application of industry standards. This means that strategic automation delivers ***ultimate flexibility through enforced rigidity*** (see Figure 20.2). Figure 20.3 outlines some key principles that must be remembered when applying automation, and Figure 20.4 shows the five phases of the strategic automation project methodology.

You can plan on spending about 50% of the engineering effort before you even begin programming the control system (see Figure 20.5). ISA's S88.01 Batch Control standard maps physical process equipment to procedures to produce process actions that in turn make product (Figure 20.6).

When modularizing a process, you should work from the top down. When building procedures, you should work from the bottom up. Remember, the great pyramids were built from the bottom up (Figure 20.7).

There are some criteria that can be used to establish the boundaries of modules. The most important thing to remember, however, is that

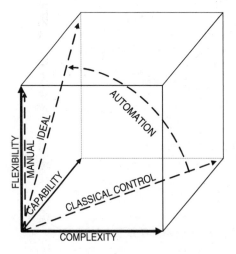

Figure 20.2 Process capability—automation.

- Know your automation objective
- Solicit your experts
- Simplify and optimize
- Focus on adding value
- Apply modular manufacturing principles
- Make your process communicate
- Define! Define! Define!
- Retrain and realign
- Stick with it until it works
- Continuous improvement

Figure 20.3 Key elements of successful automation.

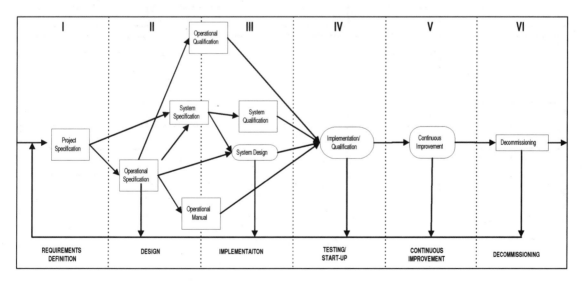

Figure 20.4 Strategic automation methodology.

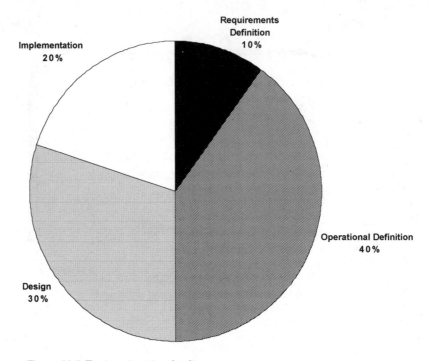

Figure 20.5 Engineering time loading.

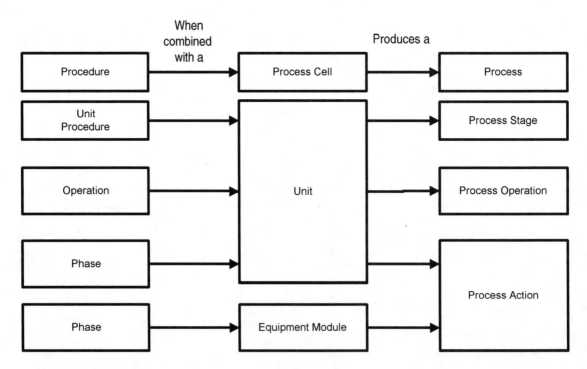

Figure 20.6 Model integration.

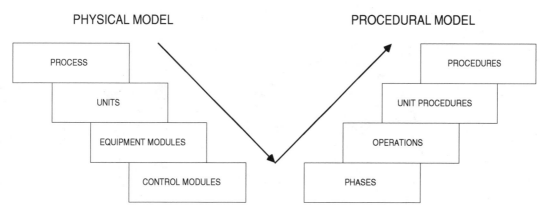

Figure 20.7 Modularization steps.

everything in the module should have the same purpose or be a system (Figures 20.8 and 20.9).

The Future

There are a few more things to keep in mind as we move forward. Things are changing very rapidly in the technology arena. We need to keep in mind things like artificial intelligence and expert systems, which can potentially help in managing the processes better. Simulation is already being used for process modeling and check-outs, and who knows how virtual reality may be used? Multimedia is becoming better, faster, and cheaper. Some are starting to integrate it into systems for training and help videos. Proprietary systems are being re-

- Purpose
- Use
- Portability
- Flexibility
- Independence
- Expansion
- Isolation
- Physical process constraints

Figure 20.8 Modularization criteria.

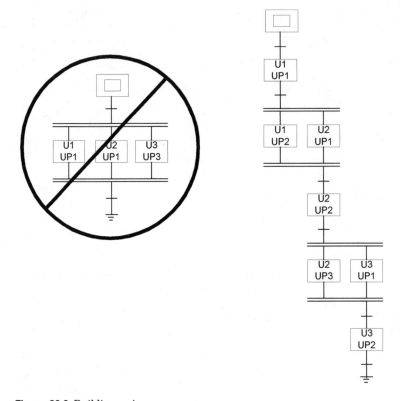

Figure 20.9 Building recipes.

placed by open, commodity systems. Could distributed control systems (DCSs) go the way of the mainframe? These are just some things to think about.

Common Traps to Avoid

The following are some common traps to *avoid* when implementing automation:

- The right people not involved
- No clearly defined objectives
- Don't allow enough time for definition
- Don't allow enough time for follow-through
- Automate current procedures
- Automate part way
- Try to start too big

Summary

VICTOR: "Strategic automation methodology was very successful when applied to our project. I think that we need to continue using it."

LARRY: "I hate to admit it, but I agree. It certainly was a smooth start-up."

Example Process

DISCLAIMER: This process is shown for illustrative purposes only. It in no way is intended to be an actual process. Any similarity to any process, real or fictional, is purely coincidental. There is no intention for this process to be able to produce any product, nor is there any intention for this process to actually be operable. This specification is focused on illustrating the operational portion of the process. In a normal project, there would be many other aspects of the control system and strategy that would need to be defined. For example, each Control Module and Equipment Module would need to be defined in more detail. The inputs, outputs, alarms, interlocks, control response, and many other things would be defined. For simplicity, we have assumed that all of the information about the actual system, equipment modules, and control modules were defined in a separate format.

Process Overview

The process consists of 13 distinct physical equipment stations. The heart of the process is the two flexible reactor units. Several raw materials and a catalyst feed the reactors. Utilities service the two reactors, including an oxygen analyzer and a vacuum system. Functionally, the reactors need to be able to agitate, heat, react, strip, transfer, purge, blanket, wash, and rinse. Many of these things need to be done at the same time.

Upstream of the reactors are the catalyst and three different Raw Materials. At this point assume RAWMAT1 will be a tank loaded from a tanker both reactors can load from at the same time. RAWMAT2 will be a bag loading station used by only one reactor at a time. RAWMAT3 will be a plant header supplying material whenever either reactor needs it. Each of the catalyst feeds is a drum loading station servicing only one reactor.

Downstream of the reactor is a batch distillation column. This can be fed by one reactor during a batch. The column will be designed flexibly enough to allow either the bottoms or the overheads to be sent to bulk storage and the rest sent to waste.

Bulk storage has some flexibility of it's own. It will allow solvent in to wash the vessels, allow recipe values to determine which tank receives the product, and allow bypass into awaiting tankers.

The thought behind the entire process is allow for absolute maximum flexibility and reusability of work already designed. It will allow one reactor to run while the other is cleaned, or, if the same product is being campaigned out of one reactor, the other can run but not have the use of the column. The process could also run both reactors directly to the bulk storage unit, using different vessels of course (see Figure A.1).

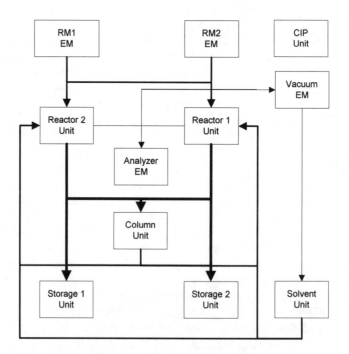

Figure A.1 Sample process block diagram.

Automation Objectives

1. Flexibility
   a. Able to make any product that the process is capable of making
   b. Able to take a product from scale-up to production in two days
2. Regulatory compliance
3. Validated
4. Safe
5. Detailed batch report produced with each batch of product

Analyzer EM Specification

Written by: _____ _____

　　　　　　Irene (Process Automation Engineer)　　　　Date

Reviewed by: _____ _____

　　　　　　Steve (Start-up Coordinator)　　　　　　　Date

_____ _____

　　　　　　Owen (Process Operator)　　　　　　　　　Date

Approved By: _____ _____

　　　　　　Patty (Process Engineer)　　　　　　　　　Date

REVISED November 22, 1996

REVISION LOG

Original:　　　Irene　　　　　　　Date: November 22, 1996

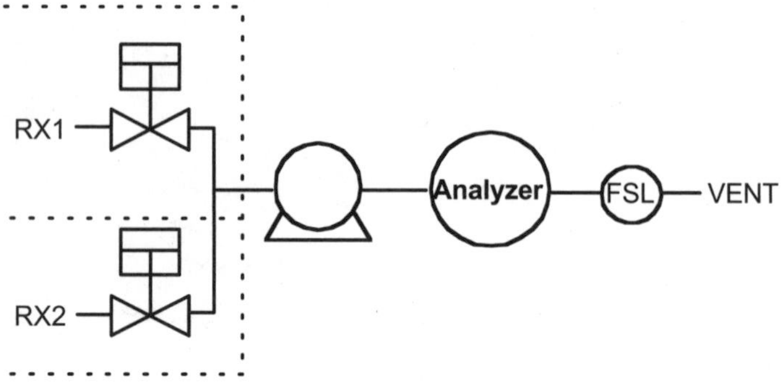

Figure A.2 Analyzer EM.

The analyzer equipment module can measure the O_2 level in either reactor, but only one at a time.

Module	Module Type	Phases
INTERNAL		
Sample Pump	CM	ANALYZE
ACQUIRED		
Sample valve to Reactor 1	CM	
Sample valve to Reactor 2	CM	

Analyze Phase

During the analyze phase the content of a tank is analyzed for percentage of oxygen.

STATE	RUNNING			STOPPING	RESTARTING	ABORTING	HOLDING
	S1 T1 S2 T2 S3 T3						
STEP	S1	S2	S3	S1	S1	S1	S1
REQUESTS							
INTERNAL MODULES							
Sample Pump CM	Start		Stop				
ACQUIRED MODULES							
Sample valve to Reactor 1 CM	***		Close	Close		Close	Close
Sample valve to Reactor 2 CM	***		Close	Close		Close	Close
ALARMS							
Low Flow	Disable	Enable	Disable	Disable	Enable	Disable	Disable
TRANSITIONS							
Unit ready to transfer and Fow > FSL	T1						
2 Minutes		T2					
Pump off			T3	Complete		Complete	Complete
MESSAGES							
Waiting to initiate sample	151						
Sampling		152					
Analysis complete			159				
PARAMETERS							
P1 Test #							
REPORTS							

*** one of the valves open based on which tank is being sampled, the other closes.

PERMISSIVES AND INTERLOCKS

CONDITION	TYPE
Flow below FSL after step S2	I

P: PERMISSIVE; I: INTERLOCK

CIP Equipment Module Specification

Written by: _____   _____
Irene (Process Automation Engineer)   Date

Reviewed by: _____   _____
Steve (Start-up Coordinator)   Date

_____   _____
Owen (Process Operator)   Date

Approved By: _____   _____
Patty (Process Engineer)   Date

REVISED November 22, 1996

REVISION LOG

Original: Irene   Date: November 22, 1996

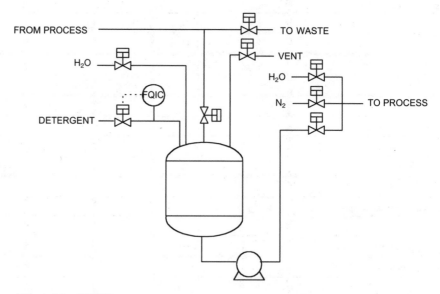

Figure A.3   CIP EM.

The CIP unit is used to clean the process equipment without having to remove it. It goes through three cycles: wash, rinse, and dry. It works in conjuction with the individual unit being cleaned.

Module	Module Type	Phases
INTERNAL		CIP
RO water to process valve	CM	
N2 valve	CM	
CIP tank outlet valve	CM	
CIP tank pump	CM	
CIP tank process return valve	CM	
Waste valve	CM	
RO water to CIP tank valve	CM	
Detergent Control Module	CM	
Vent valve	CM	
ACQUIRED		

CIP Phase

During the CIP phase, the CIP unit is taken through the wash, rinse, and dry cycles to clean some pieces of equipment. The unit being cleaned has a phase that coordinates with the CIP phase.

STATE	RUNNING							STOPPING	RESTARTING	ABORTING	HOLDING
STEP	S1	S2	S3	S4	S5	S6	S7	S1	S1	S1	S1
REQUESTS											
Ready to Clean unit	10										
Done cleaning unit							11				
INTERNAL MODULES											
RO water to process valve CM				Open	Close			Close	-	Close	Close
N2 valve CM				Open			Close	Close	-	Close	Close
CIP tank outlet valve CM		Open	Close					Close	-	Close	Close
CIP tank pump CM		Start	Stop					Stop	-	Stop	Stop
CIP tank process return valve CM				**		Close		Close	-	Close	Close
Waste valve CM		Open	Close	***		Close		Close	-	Close	Close
RO water to CIP tank valve CM		***	Close					Close	-	Close	Close
Detergent Control Module CM	Auto- SP							Close	-	Close	Close
Vent valve CM					Open		Close	Close	-	Close	Close
EXTERNAL MODULES											
ALARMS											
TRANSITIONS											
Unit ready	T1										
Time > P1		T2									
Valves closed			T3				T7	Complete		Complete	Complete
Time > P2				T4							

STATE	RUNNING							STOPPING	RESTARTING	ABORTING	HOLDING
STEP	S1	S2	S3	S4	S5	S6	S7	S1	S1	S1	S1
Time > (time to drain lines)					T5						
Time > Dry time						T6					
MESSAGES											
PARAMETERS											
P1 Time to wash											
P2 Time to Rinse											
P3 Time to Dry											
REPORTS											
R1 Actual Wash time											
R2 Actual Rinse time											
R3 Actual Dry time											

(Diagram above RUNNING column: S1—T1, S2—T2, S1—T3, S1—T4, S1—T5, S1—T6, S1—T7)

*   Open if Level in CIP tank goes below 10%
**   Open until the Level in CIP tank goes above 80%
***   Closed until the Level in CIP tank goes above 80%

PERMISSIVES AND INTERLOCKS

CONDITION	TYPE

P:PERMISSIVE ; I:INTERLOCK

Distillation Column Unit Specification

Written by: _____   _____

Irene (Process Automation Engineer)   Date

Reviewed by: _____   _____

Steve (Start-up Coordinator)   Date

_____   _____

Owen (Process Operator)   Date

Approved By: _____   _____

Patty (Process Engineer)   Date

REVISED November 22, 1996

REVISION LOG

Original: Irene   Date: November 22, 1996

The Distillation column unit is used to separate materials. It is a flexible batch distillation unit. The overheads and bottoms can each be sent to either reactor or storage tank.

Module(s)	Module Type	Phase(s)
INTERNAL		Fill_Bottoms
Bottoms Fill Valve	CM	Fill_Overheads
Bottoms Level Controller	CM	Reflux
Bottoms Pump	CM	Reboil
Overheads Fill Valve	CM	Distil
Reflux Flow Controller	CM	
Overheads Outlet Valve	CM	
Overheads Level Controller	CM	
Condenser Cooling H2O	CM	
Reboil Steam controller	CM	
Preheater Steam Valve	CM	
Distillation Column Transfer	EM	Overheads_To_Rxtr1
		Overheads_To_Rxtr2
		Overheads_To_Tank1
		Overheads_To_Tank2
		Bottoms_To_Rxtr1
		Bottoms_To_Rxtr2
		Bottoms_To_Tank1
		Bottoms_To_Tank2
ACQUIRED		

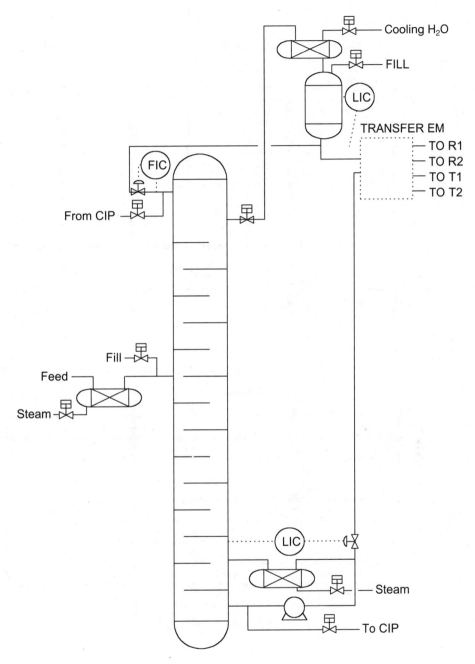

Figure A.4 Distillation column unit.

Fill Bottoms Phase

During the fill_bottoms phase the bottom of the column is filled with the appropriate liquid.

	S1	S2	STOPPING	RESTARTING	ABORTING	HOLDING
STATE	RUNNING		STOPPING	RESTARTING	ABORTING	HOLDING
STEP	S1	S2	S1	S1	S1	S1
REQUESTS						
INTERNAL MODULES						
Bottoms Fill Valve	Open	Close	Close	Open	Close	Close
Bottoms Level Controller	Manual-0.0	-				
Bottoms Pump	Stop	-				
ALARMS						
TRANSITIONS						
Level > P1	T1					
Valve closed		T2	Complete		Complete	Complete
MESSAGES						
PARAMETERS						
P1 Level to fill bottoms						
REPORTS						
R1 Actual level						

Diagram: S1, T1, S2, T2

PERMISSIVES AND INTERLOCKS

CONDITION	TYPE

P: PERMISSIVE; I: INTERLOCK

Fill Overheads Phase

During the fill_overheads phase the column overheads recovery tank is filled with the appropriate liquid.

		RUNNING	STOPPING	RESTARTING	ABORTING	HOLDING
STATE S1 ┬T1 S2 ┬T2						
STEP	S1	S2	S1	S1	S1	S1
REQUESTS						
INTERNAL MODULES						
Overheads Fill Valve	Open	Close	Close		-	Close	Close
Reflux Flow Controller	Manual-0.0					
Overheads Outlet Valve	Close					
Overheads Level Conttroller	Manual-0.0					
ALARMS						
TRANSITIONS						
Level > P1	T1					
Valve closed		T2	Complete		Complete	Complete
MESSAGES						
PARAMETERS						
P1 Level sestpoint						
REPORTS						
R1 Actual level						

PERMISSIVES AND INTERLOCKS

CONDITION	TYPE

P: PERMISSIVE; I: INTERLOCK

Reflux Phase

During the reflux phase the overheads liquid is placed back into the distillation column.

	STATE	RUNNING	STOPPING	RESTARTING	ABORTING	HOLDING
	STEP	S1	S1	S1	S1	S1
REQUESTS						
INTERNAL MODULES						
Reflux Flow Controller		Auto-P1	Manual-0.0	-	Manual-0.0	Manual-0.0
Condenser Cooling H2O		Open	Close	-	Close	-
Overheads Outlet Valve		Open	Close	-	Close	-
ALARMS						
TRANSITIONS						
Valves closed			Complete		Complete	Complete
MESSAGES						
PARAMETERS						
P1 Reflux flow setpoint						
REPORTS						

PERMISSIVES AND INTERLOCKS

CONDITION	TYPE

P: PERMISSIVE; I: INTERLOCK

Reboil Phase

During the reboil phase the bottoms liquid is recycled and heated.

STATE	RUNNING	STOPPING	RESTARTING	ABORTING	HOLDING
STEP	S1	S1	S1	S1	S1
REQUESTS					
INTERNAL MODULES					
Bottoms Pump	Start	Stop		Stop	Stop
Reboil Steam controller	Open	Close		Close	Close
ALARMS					
TRANSITIONS					
Pump stopped		Complete		Complete	Complete
MESSAGES					
PARAMETERS					
REPORTS					

PERMISSIVES AND INTERLOCKS

CONDITION	TYPE

P: PERMISSIVE; I: INTERLOCK

Distill Phase

During the distill phase, the distillation column is placed on-line, receiving material from upstream, and sending both overheads and bottoms material downstream.

STATE	RUNNING	STOPPING	RESTARTING	ABORTING	HOLDING
STEP	S1	S1	S1	S1	S1
REQUESTS					
INTERNAL MODULES					
Preheater Steam Valve	Open	Close		Close	Close
Bottoms Level Control	Auto-P1	Manual-0.0		Manual-0.0	Manual-0.0
ALARMS					
TRANSITIONS					
Valve closed		Complete		Complete	Complete
MESSAGES					
PARAMETERS					
P1 Bottoms Level control Setpoint					
P2 – Distillation temperature					
REPORTS					

PERMISSIVES AND INTERLOCKS

CONDITION	TYPE

P: PERMISSIVE; I: INTERLOCK

Clean Phase

During the clean phase, the distillation column is cleaned by the CIP system. The clean phase coordinates with the CIP plase to open and close the valves.

	S1 T1 S2 T2	RUNNING	STOPPING	RESTARTING	ABORTING	HOLDING
STATE		RUNNING	STOPPING	RESTARTING	ABORTING	HOLDING
STEP	S1	S2	S1	S1	S1	S1
REQUESTS						
Ready to clean	10					
Wait for done cleaning message		11				
INTERNAL MODULES						
CIP inlet valve	Open	Close	Close		Close	Close
CIP return valve	Open	Close	Close		Close	Close
EXTERNAL MODULES						
ALARMS						
TRANSITIONS						
CIP system ready	T1					
CIP system done		T2				
MESSAGES						
PARAMETERS						
P1 Temperature setpoint						
P2 Time to hold temperature						
REPORTS						

PERMISSIVES AND INTERLOCKS

CONDITION	TYPE

P: PERMISSIVE; I: INTERLOCK

Distillation Column Transfer
Equipment Module

The distillation column transfer equipment module (Figure A.5) is used to transfer the material from both the overheads and bottoms of the column to one of the reactors or storage tanks.

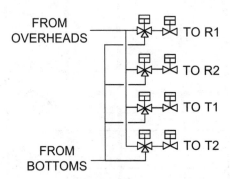

Figure A.5   Distillation column transfer EM.

Overheads_to_RXTR1 Phase

During the overheads_to_RXTR1 phase, material is transferred to one of the reactors.

STATE	\	RUNNING	\	STOPPING	RESTARTING	ABORTING	HOLDING
STEP	S1	S2	S3	S1	S1	S1	S1
REQUESTS							
Ready to transfer	01						
Waiting for complete message		02					
Report total loaded			X				
INTERNAL MODULES							
Reactor 1 Feed Valve	Open		Close	Close	Open	Close	Close
Reactor 1 3-way valve	Ovrhds						
ALARMS							
TRANSITIONS							
Reactor valves open	T1						
Transfer complete		T2					
Valves closed			T3	Complete		Complete	Complete
MESSAGES							
Opening reactor valves	151						
Transferring		152					
Transfer complete			159				
PARAMETERS							
REPORTS							

The state diagram shows: S1 — T1 — S2 — T2 — S3 — T3

PERMISSIVES AND INTERLOCKS

CONDITION	TYPE

P: PERMISSIVE; I: INTERLOCK

Overheads_to_RXTR2 Phase

During the overheads_to_RXTR2 phase, material is transferred to one of the reactors.

	RUNNING S1	RUNNING S2	RUNNING S3	STOPPING S1	RESTARTING S1	ABORTING S1	HOLDING S1
REQUESTS							
Ready to transfer	01						
Waiting for complete message		02					
Report total loaded			X				
INTERNAL MODULES							
Reactor 2 Feed Valve	Open		Close	Close	Open	Close	Close
Reactor 2 3-way valve	Ovrhds						
ALARMS							
TRANSITIONS							
Reactor valves open	T1						
Transfer complete		T2					
Valves closed			T3	Complete		Complete	Complete
MESSAGES							
Opening reactor valves	151						
Transferring		152					
Transfer complete			159				
PARAMETERS							
REPORTS							

PERMISSIVES AND INTERLOCKS

CONDITION	TYPE

P: PERMISSIVE; I: INTERLOCK

Overheads_to_Tank1 Phase

During the overheads_to_tank 2 phase, material is transferred to one of the reactors.

	S1	S2	S3 (RUNNING)	STOPPING	RESTARTING	ABORTING	HOLDING
STATE			RUNNING	STOPPING	RESTARTING	ABORTING	HOLDING
STEP	S1	S2	S3	S1	S1	S1	S1
REQUESTS							
Ready to transfer	01						
Waiting for complete message		02					
Report total loaded			X				
INTERNAL MODULES							
Storage Tank 1 Feed Valve	Open		Close	Close	Open	Close	Close
Storage Tank 1 3-way valve	Ovrhds						
Storage Tank 2 3-way valve CM							
ALARMS							
TRANSITIONS							
Tank valves open	T1						
Transfer complete		T2					
Valves closed			T3	Complete		Complete	Complete
MESSAGES							
Opening Tank valves	151						
Transferring		152					
Transfer complete			159				
PARAMETERS							
REPORTS							

PERMISSIVES AND INTERLOCKS

CONDITION	TYPE

P: PERMISSIVE; I: INTERLOCK

Overheads_to_Tank 2 Phase

During the overheads_to_tank 2 phase, material is transferred to one of the reactors.

	RUNNING			STOPPING	RESTARTING	ABORTING	HOLDING
STATE							
STEP	S1	S2	S3	S1	S1	S1	S1
REQUESTS							
Ready to transfer	01						
Waiting for complete message		02					
Report total loaded			X				
INTERNAL MODULES							
Storage Tank 2 Feed Valve	Open		Close	Close	Open	Close	Close
Storage Tank 2 3-way valve	Ovrhds						
ALARMS							
TRANSITIONS							
Tank valves open	T1						
Transfer complete		T2					
Valves closed			T3	Complete		Complete	Complete
MESSAGES							
Opening Tank valves	151						
Transferring		152					
Transfer complete			159				
PARAMETERS							
REPORTS							

State diagram (RUNNING column): S1 — T1 — S2 — T2 — S3 — T3

PERMISSIVES AND INTERLOCKS

CONDITION	TYPE

P: PERMISSIVE; I: INTERLOCK

Bottoms_to_RXTR1 Phase

During the bottoms_to_RXTR1 phase, material is transferred to one of the reactors.

	RUNNING S1	RUNNING S2	RUNNING S3	STOPPING S1	RESTARTING S1	ABORTING S1	HOLDING S1
REQUESTS							
Ready to transfer	01						
Waiting for complete message		02					
Report total loaded			X				
INTERNAL MODULES							
Reactor 1 Feed Valve	Open		Close	Close	Open	Close	Close
Reactor 1 3-way valve	Ovrhds						
ALARMS							
TRANSITIONS							
Reactor valves open	T1						
Transfer complete		T2					
Valves closed			T3	Complete		Complete	Complete
MESSAGES							
Opening reactor valves	151						
Transferring		152					
Transfer complete			159				
PARAMETERS							
REPORTS							

State diagram (STATE column):
S1
— T1
S2
— T2
S3
— T3

PERMISSIVES AND INTERLOCKS

CONDITION	TYPE

P: PERMISSIVE; I: INTERLOCK

Bottoms_to_RXTR2 Phase

During the bottoms_to_RXTR2 phase, material is transferred to one of the reactors.

STATE / STEP	RUNNING S1	RUNNING S2	RUNNING S3	STOPPING S1	RESTARTING S1	ABORTING S1	HOLDING S1
REQUESTS							
Ready to transfer	01						
Waiting for complete message		02					
Report total loaded			X				
INTERNAL MODULES							
Reactor 2 Feed Valve	Open		Close	Close	Open	Close	Close
Reactor 2 3-way valve	Ovrhds						
ALARMS							
TRANSITIONS							
Reactor valves open	T1						
Transfer complete		T2					
Valves closed			T3	Complete		Complete	Complete
MESSAGES							
Opening reactor valves	151						
Transferring		152					
Transfer complete			159				
PARAMETERS							
REPORTS							

State diagram: S1 — T1 — S2 — T2 — S3 — T3

PERMISSIVES AND INTERLOCKS

CONDITION	TYPE

P: PERMISSIVE; I: INTERLOCK

Bottoms_to_Tank1 Phase

During the bottoms_to_tank1 phase, material is transferred to one of the reactors.

STATE / STEP	S1	S2	S3 (RUNNING)	STOPPING S1	RESTARTING S1	ABORTING S1	HOLDING S1
REQUESTS							
Ready to transfer	01						
Waiting for complete message		02					
Report total loaded			X				
INTERNAL MODULES							
Storage Tank 1 Feed Valve	Open		Close	Close	Open	Close	Close
Storage Tank 1 3-way valve	Ovrhds						
ALARMS							
TRANSITIONS							
Tank valves open	T1						
Transfer complete		T2					
Valves closed			T3	Complete		Complete	Complete
MESSAGES							
Opening Tank valves	151						
Transferring		152					
Transfer complete			159				
PARAMETERS							
REPORTS							

The STATE diagram shows: S1 — T1 — S2 — T2 — S3 — T3 (ground).

PERMISSIVES AND INTERLOCKS

CONDITION	TYPE

P: PERMISSIVE; I: INTERLOCK

Bottoms_to_Tank2 Phase

During the bottoms_to_tank2 phase, material is transferred to one of the reactors.

STATE			RUNNING	STOPPING	RESTARTING	ABORTING	HOLDING
STEP	S1	S2	S3	S1	S1	S1	S1
REQUESTS							
Ready to transfer	01						
Waiting for complete message		02					
Report total loaded			X				
INTERNAL MODULES							
Storage Tank 2 Feed Valve	Open		Close	Close	Open	Close	Close
Storage Tank 2 3-way valve	Ovrhds						
ALARMS							
TRANSITIONS							
Tank valves open	T1						
Transfer complete		T2					
Valves closed			T3				
MESSAGES							
Opening Tank valves	151						
Transferring		152					
Transfer complete			159	Complete		Complete	Complete
PARAMETERS							
REPORTS							

S1
—|— T1
S2
—|— T2
S3
—|— T3

PERMISSIVES AND INTERLOCKS

CONDITION	TYPE

P: PERMISSIVE; I: INTERLOCK

Raw Material 1 EM Specification

Written by: _____ _____

Irene (Process Automation Engineer)   Date

Reviewed by: _____ _____

Steve (Start-up Coordinator)   Date

_____ _____

Owen (Process Operator)   Date

Approved By: _____ _____

Patty (Process Engineer)   Date

REVISED November 22, 1996

REVISION LOG

Original: Irene   Date: November 22, 1996

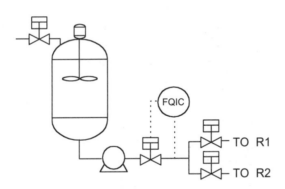

Figure A.6   Raw material 1 EM.

The raw material 1 equipment module can feed either reactor, but only one at a time.

Module(s)	Module Type	Phase(s)
INTERNAL		FILL
Fill valve	CM	FEED_RM1
Raw Material pump	CM	
Feed valve	CM	
Bulk storage agitator	CM	
ACQUIRED		
R1 feed Valve	CM	
R2 feed Valve	CM	

Fill Phase

During the fill phase, material is loaded into the raw material tank.

STATE	RUNNING	STOPPING	RESTARTING	ABORTING	HOLDING
STEP	S1	S1	S1	S1	S1
REQUESTS					
INTERNAL MODULES					
Fill valve	Open	Close	-	Close	Close
ALARMS					
TRANSITIONS					
Valves closed		Complete		Complete	Complete
MESSAGES					
Ready to be filled	151				
PARAMETERS					
REPORTS					

PERMISSIVES AND INTERLOCKS

CONDITION	TYPE
Not HIHI level in tank	PI

P: PERMISSIVE; I: INTERLOCK

Feed_RM1 Phase

During the Feed_RM1 phase, material is transferred to one of the reactors.

STATE	RUNNING			STOPPING	RESTARTING	ABORTING	HOLDING
STEP	S1	S2	S3	S1	S1	S1	S1
REQUESTS							
INTERNAL MODULES							
Raw Material pump	Start		Stop	Stop		Stop	
Feed valve		Auto-P1					
Bulk storage agitator	Start		Stop	Stop		Stop	
R1 feed Valve		*		Close	*	Close	Close
R2 feed Valve		*		Close	*	Close	Close
ALARMS							
TRANSITIONS							
Unit ready to transfer	T1						
Transfer complete		T2					
Valve closed			T3	Complete		Complete	Complete
MESSAGES							
Waiting to initiate transfer	151						
Transferring		152					
Transfer complete			159				
PARAMETERS							
P1 Amount to transfer							
REPORTS							

* Open based on which reactor unit owns the EM

PERMISSIVES AND INTERLOCKS

CONDITION	TYPE

P: PERMISSIVE; I: INTERLOCK

Raw Material 2 EM Specification

Written by: _____   _____
　　　　　　　Irene (Process Automation Engineer)　　　Date

Reviewed by: _____   _____
　　　　　　　Steve (Start-up Coordinator)　　　　　Date

　　　　　　　_____   _____
　　　　　　　Owen (Process Operator)　　　　　　Date

Approved By: _____   _____
　　　　　　　Patty (Process Engineer)　　　　　Date

REVISED November 22, 1996

REVISION LOG

Original:　　Irene　　　　　　Date: November 22, 1996

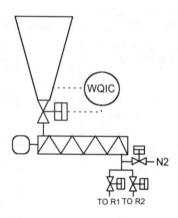

Figure A.7　Raw material 2 EM.

The raw material 2 equipment module can feed either reactor, but only one at a time. It is a bag dump station for transferring dry materials into the reactor. It is first loaded with material and then blows the material and controls the total transferred.

Module(s)	Module Type	Phase(s)
INTERNAL		FILL LOAD_RM2
Bottom valve	CM	
Auger motor	CM	
Nitrogen valve	CM	
ACQUIRED		
R1 feed Valve	CM	
R2 feed Valve	CM	

Fill Phase

During the fill phase, material is manually loaded into the hopper.

STATE	RUNNING	STOPPING	RESTARTING	ABORTING	HOLDING
STEP	S1	S1	S1	S1	S1
REQUESTS					
INTERNAL MODULES					
Bottom valve	Close	-	-	-	-
Auger motor	Stop	-	-	-	-
Nitrogen valve	Close	-	-	-	-
ALARMS					
TRANSITIONS					
MESSAGES					
Ready to fill hopper	151				
PARAMETERS					
REPORTS					

PERMISSIVES AND INTERLOCKS

CONDITION	TYPE
Hopper weight not above WAHH	PI

P: PERMISSIVE; I: INTERLOCK

STATE	RUNNING			STOPPING	RESTARTING	ABORTING	HOLDING
STEP	S1	S2	S3	S1	S1	S1	S1
REQUESTS							
Ready to transfer	01						
Waiting for stop		02					
Report total loaded			X				
INTERNAL MODULES							
Bottom valve		Open	Close	Close		Close	
Auger motor		Start	Stop	Stop	Start	Stop	Stop
Nitrogen valve	Open		Open	Close	Open	Close	Close
R1 feed Valve		*		Close	*	Close	Close
R2 feed Valve		*		Close	*	Close	Close
ALARMS							
TRANSITIONS							
Unit ready to transfer	T1						
Transfer complete		T2					
Valve closed			T3	Complete		Complete	Complete
MESSAGES							
Waiting to initiate transfer	151						
Transferring		152					
Transfer complete			159				
PARAMETERS							
P1 Amount to transfer							
REPORTS							

In the STATE cell the following SFC diagram is drawn:

S1
— T1
S2
— T2 T2
S3
— T3

PERMISSIVES AND INTERLOCKS

CONDITION	TYPE

P: PERMISSIVE; I: INTERLOCK

Reactor Unit Specification

Written by: _____ _____
Irene (Process Automation Engineer)   Date

Reviewed by: _____ _____
Steve (Start-up Coordinator)   Date

_____ _____
Owen (Process Operator)   Date

Approved By: _____ _____
Patty (Process Engineer)   Date

REVISED November 22, 1996

REVISION LOG

Original: Irene   Date: November 22, 1996

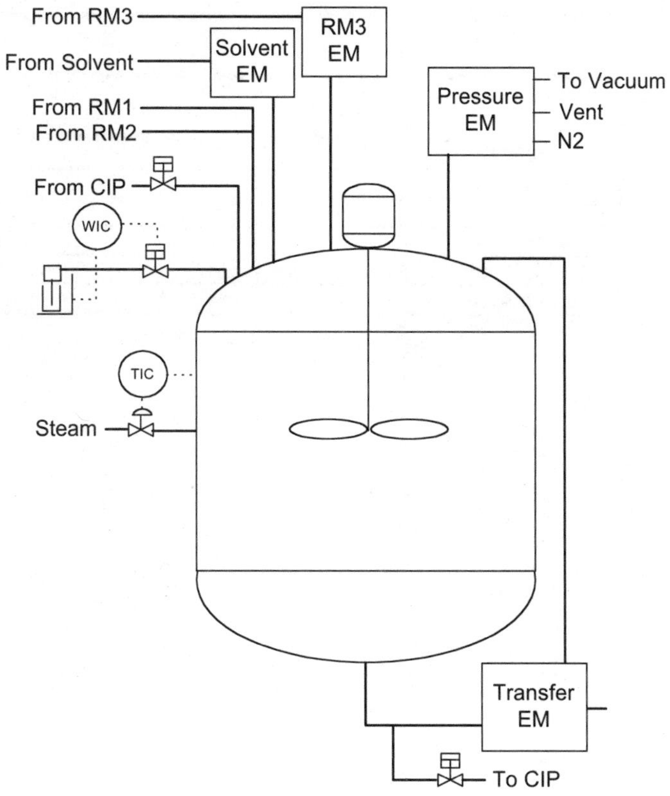

Figure A.8 Reactor unit.

The heart of the process is the two flexible reactor units. The reactors are fed by several raw materials and a catalyst. Utilities service the two reactors, including an oxygen analyzer and a vacuum system. Functionally, each reactor can agitate, heat, react, strip, transfer, purge, blanket, wash, and rinse. Many of these things need to be done at the same time.

Module(s)	Module Type	Phase(s)
INTERNAL		Mix Heat Clean
Agitator	CM	
Temperature controller	CM	
CIP inlet valve	CM	
CIP return valve	CM	
Raw material #3	EM	Load_Mat3
Solvent	EM	Load_Solvent
Transfer	EM	XFER_OUT Recycle
Pressure	EM	Strip Pressure_Ctl
ACQUIRED		
Analyzer	EM	Analyze
CIP	EM	CIP
Raw Material 1	EM	Feed_RM1
Raw Material 2	EM	Feed_RM2
Vacuum system	EM	On-Line

Mix Phase

The mix phase turns on the agitator, and runs until it receives a command to stop.

	RUNNING	STOPPING	RESTARTING	ABORTING	HOLDING
STATE	RUNNING	STOPPING	RESTARTING	ABORTING	HOLDING
STEP	S1	S1	S1	S1	S1
REQUESTS					
INTERNAL MODULES					
Agitator	Start	Stop	-	Stop	Stop
EXTERNAL MODULES					
ALARMS					
TRANSITIONS					
(Time>P1) and (P1>0)	T1				
Agitator stopped		Complete		Complete	Complete
MESSAGES					
PARAMETERS					
P1-Time to Mix					
REPORTS					
R1-Actual time mixed					

The state diagram in the RUNNING column shows:

S1 → T1 → S2 → T2 → (ground)

PERMISSIVES AND INTERLOCKS

CONDITION	TYPE

P: PERMISSIVE; I: INTERLOCK

Heat Phase

During the heat phase the reactor is heated using steam to a setpoint temperature. The temperature is then maintained for a specified time. If a time is not specified (P2 = 0), the temperature is maintained until the phase is stopped.

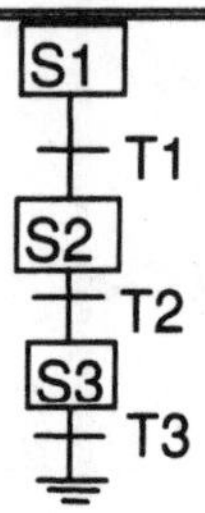
S1
T1
S2
T2
S3
T3

STATE	RUNNING			STOPPING	RESTARTING	ABORTING	HOLDING
STEP	S1	S2	S3	S1	S1	S1	S1
REQUESTS							
INTERNAL MODULES							
Temperature controller	Auto-P1		0.0	0.0	Auto-P1	0.0	0.0
EXTERNAL MODULES							
ALARMS							
TRANSITIONS							
Temperature > P1	T1						
(P2 > 0) and (Temperature > P2)		T2					
Valve closed			T3	Complete		Complete	Complete
MESSAGES							
PARAMETERS							
P1 Temperature setpoint							
P2 Time to hold temperature							
REPORTS							

PERMISSIVES AND INTERLOCKS

CONDITION	TYPE

P: PERMISSIVE I: INTERLOCK

Clean Phase

During the clean phase the reactor is cleaned by the CIP system. The clean phase coordinates with the CIP phase to open and close the valves.

	RUNNING		STOPPING	RESTARTING	ABORTING	HOLDING
	S1 — T1 S2 — T2					
STATE	RUNNING		STOPPING	RESTARTING	ABORTING	HOLDING
STEP	S1	S2	S1	S1	S1	S1
REQUESTS						
Ready to clean	10					
Wait for done cleaning message		11				
INTERNAL MODULES						
CIP inlet valve	Open	Close	Close		Close	Close
CIP return valve	Open	Close	Close		Close	Close
EXTERNAL MODULES						
ALARMS						
TRANSITIONS						
CIP system ready	T1					
CIP system done		T2				
MESSAGES						
PARAMETERS						
P1 Temperature setpoint						
P2 Time to hold temperature						
REPORTS						

PERMISSIVES AND INTERLOCKS

CONDITION	TYPE

P: PERMISSIVE; I: INTERLOCK

Equipment Modules

The raw material # 3 equipment module controls the charging of raw material 3 from a plant service header into the reaction vessel.

Raw Material #3

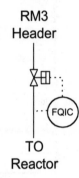

Figure A.9   Reactor RM3 EM.

Load_mat3 Phase

During the load_mat3 phase, a specified amount of raw material #3 is loaded into the reactor.

STATE	RUNNING		STOPPING	RESTARTING	ABORTING	HOLDING
STEP	S1	S2	S1	S1	S1	S1
REQUESTS						
INTERNAL MODULES						
Feed valve	Open	Close	Close	-	Close	Close
EXTERNAL MODULES						
ALARMS						
TRANSITIONS						
Amount loaded > P1	T1					
Valve closed		T2	Complete		Complete	Complete
MESSAGES						
PARAMETERS						
P1 Amount to load						
REPORTS						

The state diagram at the top of the RUNNING column shows:

S1 → T1 → S2 → T2

PERMISSIVES AND INTERLOCKS

CONDITION	TYPE

P: PERMISSIVE; I: INTERLOCK

Solvent Equipment Module

The solvent equipment module controls the charging of solvent from the solvent storage tank into the reactor vessel.

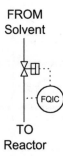

Figure A.10   Reactor solvent EM.

Load_solvent Phase

During the load_solvent phase, a specified amount of solvent is loaded into the reactor.

STATE	RUNNING		STOPPING	RESTARTING	ABORTING	HOLDING
STEP	S1	S2	S1	S1	S1	S1
REQUESTS						
INTERNAL MODULES						
Feed valve	Open	Close	Close	-	Close	Close
EXTERNAL MODULES						
ALARMS						
TRANSITIONS						
Amount loaded > P1	T1					
Valve closed		T2	Complete		Complete	Complete
MESSAGES						
PARAMETERS						
P1 Amount to load						
REPORTS						

(Diagram in RUNNING column: S1 — T1 — S2 — T2)

PERMISSIVES AND INTERLOCKS

CONDITION	TYPE

P: PERMISSIVE; I: INTERLOCK

Transfer Equipment Module

The transfer equipment module controls the phases necessary to transfer material out of and recycle it back to the reactor vessel.

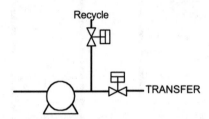

Figure A.11   Reactor transfer EM.

Xfer_out Phase

During the xfer_out phase the reactor contents are transfered to a corresponding xfer_in in another unit.

STATE	RUNNING			STOPPING	RESTARTING	ABORTING	HOLDING
STEP	S1	S2	S3	S1	S1	S1	S1
REQUESTS							
Unit ready to receive material	01						
Transfer complete			02				
INTERNAL MODULES							
Pump	-	Start	Stop	Stop	-	Stop	Stop
Discharge Valve	-	Open	Close	Close	-	Close	Close
EXTERNAL MODULES							
ALARMS							
TRANSITIONS							
Unit ready to receive material	T1						
Amount transferred > P1		T2					
Valve closed and Pump off			T3	Complete		Complete	Complete
MESSAGES							
PARAMETERS							
P1 Amount to transfer							
REPORTS							
R1 Actual amount transferred							

The sequence diagram within the table shows: S1, T1, S2, T2, S3, T3.

PERMISSIVES AND INTERLOCKS

CONDITION	TYPE

P: PERMISSIVE; I: INTERLOCK

Recycle Phase

During the recycle phase the reactor xxxxxxx xxxxxxx xxxxxxx xxxxxx xxxxx

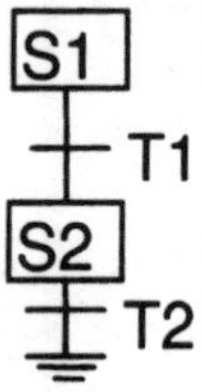

STATE	RUNNING		STOPPING	RESTARTING	ABORTING	HOLDING
STEP	S1	S2	S1	S1	S1	S1
REQUESTS						
INTERNAL MODULES						
Pump	Start	Stop	Stop	-	Stop	Stop
Recycle Valve	Open	Close	Close	-	Close	Close
EXTERNAL MODULES						
ALARMS						
TRANSITIONS						
(P1 > 0) and (time > T1)	T1					
Pump stopped and valve closed		T2	Complete		Complete	Complete
MESSAGES						
PARAMETERS						
P1 Time to recycle						
REPORTS						

The RUNNING column contains the following SFC diagram:

S1 — T1 — S2 — T2

PERMISSIVES AND INTERLOCKS

CONDITION	TYPE

P: PERMISSIVE; I: INTERLOCK

Pressure

The reactor pressure equipment module controls the phases necessary to control the pressure of the reactor.

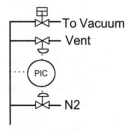

To Vacuum

Vent

PIC

N2

Figure A.12 Reactor pressure EM.

Strip Phase

During the strip phase, the vacuum system is used to pull the solvent volatiles out of the reactor.

	S1 ┬ T1 S2 ┴ T2					
STATE	RUNNING		STOPPING	RESTARTING	ABORTING	HOLDING
STEP	S1	S2	S1	S1	S1	S1
REQUESTS						
INTERNAL MODULES						
Vacuum valve	Open	Close	Close	-	Close	Close
Pressure Controller	0.0	-	-	-	-	-
EXTERNAL MODULES						
ALARMS						
TRANSITIONS						
Time > P1	T1					
Valve Closed		T2	Complete		Complete	Complete
MESSAGES						
PARAMETERS						
P1 Time to strip						
REPORTS						

PERMISSIVES AND INTERLOCKS

CONDITION	TYPE
Vacuum system must be ON-LINE	PI

P: PERMISSIVE; I: INTERLOCK

Pressure_CTL Phase

During the pressure_CTL phase, the reactor pressure is maintained at a setpoint with N_2 and the vent. The pressure is then maintained for a specified time. If a time is not specified (P2 = 0), the pressure is maintained until the phase is stopped.

	RUNNING			STOPPING	RESTARTING	ABORTING	HOLDING
STATE							
STEP	S1	S2	S3	S1	S1	S1	S1
REQUESTS							
INTERNAL MODULES							
Vacuum valve	Close	-		-	-	-	-
Pressure Controller	Auto-P1	-	0.0	0.0	-	0.0	0.0
EXTERNAL MODULES							
ALARMS							
TRANSITIONS							
Pressure > SP	T1						
(P2 > 0) and (pressure > P2)		T2					
Valve closed			T3	Complete		Complete	Complete
MESSAGES							
PARAMETERS							
P1 Pressure setpoint							
P2 Time to hold pressure							
REPORTS							
R1 Actual Pressure							
R2 Actual time							

(In the RUNNING column: S1 → T1 → S2 → T2 → S3 → T3)

PERMISSIVES AND INTERLOCKS

CONDITION	TYPE

P: PERMISSIVE; I: INTERLOCK

Storage Tank Unit Specification

Written by: _____ _____

Irene (Process Automation Engineer)   Date

Reviewed by: _____ _____

Steve (Start-up Coordinator)   Date

_____ _____

Owen (Process Operator)   Date

Approved By: _____ _____

Patty (Process Engineer)   Date

REVISED November 22, 1996

REVISION LOG

Original: Irene   Date: November 22, 1996

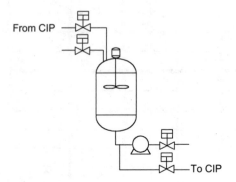

Figure A.13   Storage unit.

The storage tank units can receive material from either reactor or from either the bottoms or the overheads of the distillation column. the material is then transferred to tankers. After each product is run, the tank and lines are cleaned via the CIP system.

Module(s)	Module Type	Phase(s)
INTERNAL		XFER_IN Dump Clean
Storage tank 1 inlet valve	CM	
Storage tank 1 agitator	CM	
Storage tank 1 pump	CM	
Storage tank 1 discharge valve	CM	
Storage tank 1 pressure control	CM	
Storage tank 1 CIP inlet valve	CM	
Storage tank 1 CIP outlet valve	CM	
ACQUIRED		

Xfer_in Phase

During the xfer_in phase, material is received from an upstream unit.

STATE	RUNNING			STOPPING	RESTARTING	ABORTING	HOLDING
STEP	S1	S2	S3	S1	S1	S1	S1
REQUESTS							
Ready to receive	01						
Waiting for stop		02					
Report total loaded			X				
INTERNAL MODULES							
Storage tank 1 inlet valve		Open	Close	Close	Open	Close	Close
Storage tank 1 agitator		Start					
ALARMS							
TRANSITIONS							
Unit ready to transfer	T1						
Transfer complete		T2					
Valve closed			T3	Complete		Complete	Complete
Valve open					Complete		
MESSAGES							
Waiting to initiate transfer	151						
Transferring		152					
Transfer complete			159				
PARAMETERS							
P1 Pressure							
REPORTS							

The RUNNING column shows the state diagram: S1 → T1 → S2 → T2 → S3 → T3.

PERMISSIVES AND INTERLOCKS

CONDITION	TYPE
Level below high high level switch	PI
Not pressure deviation	I

P: PERMISSIVE; I: INTERLOCK

Dump Phase

During the dump phase, material is moved out to a tanker.

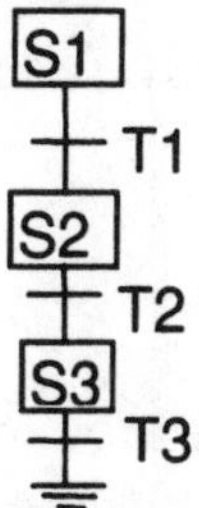

		RUNNING			STOPPING	RESTARTING	ABORTING	HOLDING
STATE								
STEP		S1	S2	S3	S1	S1	S1	S1
REQUESTS								
INTERNAL ENTITIE								
Storage tank 1 pump			Start	Stop	Stop	Start	Stop	Stop
Storage tank 1 discharge valve			Open	Close	Close	Open	Close	Close
Storage tank 1 agitator			Stop					
ALARMS								
TRANSITIONS								
Operator confirmation		T1						
Tank empty			T2					
Valve closed				T3	Complete		Complete	Complete
Valve open						Complete		
MESSAGES								
Confirm when ready for transfer		151						
Transferring			152					
Transfer complete				159				
PARAMETERS								
REPORTS								

Diagram in RUNNING column:
S1
| T1
S2
| T2
S3
| T3

PERMISSIVES AND INTERLOCKS

CONDITION	TYPE
Level below high high level switch	PI
Not pressure deviation	I

P: PERMISSIVE; I: INTERLOCK

Clean Phase

During the clean phase, the tank is cleaned using the CIP system.

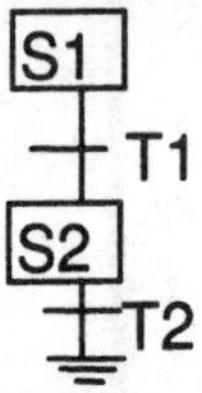

STATE	RUNNING		STOPPING	RESTARTING	ABORTING	HOLDING
STEP	S1		S1	S1	S1	S1
REQUESTS						
Ready for clean	10					
INTERNAL MODULES						
Storage tank 1 CIP inlet valve	Open	Close	Close	-	Close	Close
Storage tank 1 CIP outlet valve	Open	Close	Close	-	Close	Close
ALARMS						
TRANSITIONS						
Clean complete message received	T1					
Valves closed		T2	Complete		Complete	Complete
MESSAGES						
PARAMETERS						
REPORTS						

State diagram shown in the RUNNING column header: S1 — T1 — S2 — T2.

PERMISSIVES AND INTERLOCKS

CONDITION	TYPE
Level below high high level switch	PI

P: PERMISSIVE; I: INTERLOCK

Solvent Storage Unit Specification

Written by: _____   _____

Irene (Process Automation Engineer)   Date

Reviewed by: _____   _____

Steve (Start-up Coordinator)   Date

_____   _____

Owen (Process Operator)   Date

Approved By: _____   _____

Patty (Process Engineer)   Date

REVISED November 22, 1996

REVISION LOG

Original: Irene   Date: November 22, 1996

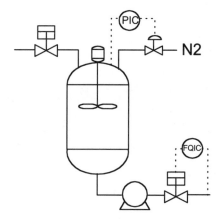

Figure A.14   Solvent unit.

The solvent storage unit can receive material from the vacuum system. The material can then either be transferred to one of the reactors or dumped to waste.

Module(s)	Module Type	Phase(s)
INTERNAL		XFER_IN XFER_OUT Pressure_Ctl
Solvent storage inlet valve	CM	
Solvent storage agitator	CM	
Solvent storage pressure control	CM	
Solvent storage pump	CM	
Solvent storage discharge valve	CM	
Solvent storage agitator	CM	
Flow Controller	CM	
ACQUIRED		

Xfer_in Phase

During the xfer_in phase, material is received from an upstream unit.

	RUNNING			STOPPING	RESTARTING	ABORTING	HOLDING
STATE	RUNNING			STOPPING	RESTARTING	ABORTING	HOLDING
STEP	S1	S2	S3	S1	S1	S1	S1
REQUESTS							
Ready to receive	01						
Waiting for stop		02					
Report total loaded			X				
INTERNAL MODULES							
Solvent storage inlet valve		Open	Close	Close	Open	Close	Close
Solvent storage agitator		Start					
Solvent storage pressure control		AUTO-p1					
ALARMS							
TRANSITIONS							
Unit ready to transfer	T1						
Transfer complete		T2					
Valve closed			T3	Complete		Complete	Complete
Valve open					Complete		
MESSAGES							
Waiting to initiate transfer	151						
Transferring		152					
Transfer complete			159				
PARAMETERS							
P1 Pressure							
REPORTS							

PERMISSIVES AND INTERLOCKS

CONDITION	TYPE
Level below high high level switch	PI
Not pressure deviation	I

P: PERMISSIVE; I: INTERLOCK

Xfer_out Phase

During the xfer_out phase, material is transferred out to a tanker.

	RUNNING			STOPPING	RESTARTING	ABORTING	HOLDING
STATE	RUNNING			STOPPING	RESTARTING	ABORTING	HOLDING
STEP	S1	S2	S3	S1	S1	S1	S1
REQUESTS							
INTERNAL MODULES							
Solvent storage pump		Start	Stop	Stop	Start	Stop	Stop
Solvent storage discharge valve		Open	Close	Close	Open	Close	Close
Solvent storage agitator		Stop					
Flow Controller		Auto-SP	0.0	0.0	Auto-SP	0.0	0.0
ALARMS							
TRANSITIONS							
Operator confirmation	T1						
FQIC setpoint reached		T2					
Valve closed			T3	Complete		Complete	Complete
Valve open					Complete		
MESSAGES							
Confirm when ready for transfer	151						
Transferring		152					
Transfer complete			159				
PARAMETERS							
REPORTS							

The step diagram shows: S1 — T1 — S2 — T2 — S3 — T3.

PERMISSIVES AND INTERLOCKS

CONDITION	TYPE
Level below high high level switch	PI
Not pressure deviation	I
Solvent Quality	P

P :PERMISSIVE; I: INTERLOCK

Pressure_ctl Phase

During the pressure_ctl phase, the tank is pressurized with nitrogen.

STATE	RUNNING	STOPPING	RESTARTING	ABORTING	HOLDING
STEP	S1	S1	S1	S1	S1
REQUESTS					
INTERNAL MODULES					
Solvent storage pressure control	Auto-P1	0.0		0.0	0.0
ALARMS					
TRANSITIONS					
Valve closed		Complete		Complete	Complete
MESSAGES					
Blanket enabled	151				
PARAMETERS					
P1 Pressure Setpoint					
REPORTS					

PERMISSIVES AND INTERLOCKS

CONDITION	TYPE
Pressure below PAHH	PI

P: PERMISSIVE; I: INTERLOCK

Vacuum System EM Specification

Written by: _____   _____

Irene (Process Automation Engineer)   Date

Reviewed by: _____   _____

Steve (Start-up Coordinator)   Date

_____   _____

Owen (Process Operator)   Date

Approved By: _____   _____

Patty (Process Engineer)   Date

REVISED November 22, 1996

REVISION LOG

Original: Irene   Date: November 22, 1996

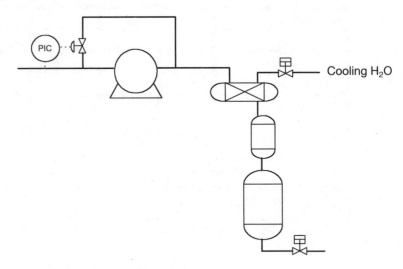

Figure A.15   Vacuum system EM.

The vacuum system unit is capable of pulling 100 inches of Hg at a volume of 1000 scfh. It can be used by either reactor unit or both at the same time.

Module(s)	Module Type	Phase(s)
INTERNAL		On-Line XFER_OUT
Vacuum pump	CM	
Vacuum recycle valve	CM	
Condenser H_2O	CM	
Accumulator outlet valve	CM	
ACQUIRED		

On-line Phase

The on-line phase starts the vacuum system and allows it to warm up before pulling vacuum.

	RUNNING		STOPPING	RESTARTING	ABORTING	HOLDING
STATE	RUNNING		STOPPING	RESTARTING	ABORTING	HOLDING
STEP	S1	S2	S1	S1	S1	S1
REQUESTS						
INTERNAL MODULES						
VACUUM PUMP	START		STOP	START	STOP	STOP
VACUUM RECYCLE VALVE	100.0	AUTO – P1	100.0	100.0	100.0	100.0
CONDENSER H2O	OPEN		CLOSE		CLOSE	
ALARMS						
TAH-1	DISABLE	ENABLE				
TAH-2	DISABLE	ENABLE				
TRANSITIONS						
20 MINUTES	T1					
MESSAGES						
WARMING UP	S1					
VACUUM SYSTEM READY		S2				
PARAMETERS						
P1 – Vacuum sepoint						
REPORTS						

The state/step column includes a diagram:

S1 — T1 — S2 — T2

PERMISSIVES AND INTERLOCKS

CONDITION	TYPE
Not high oil temperature	PI

P: PERMISSIVE; I: INTERLOCK

Xfer_out Phase

The xfer_out phase starts the vacuum system and allows it to warm up before pulling vacuum.

State diagram: S1 — T1 — S2 — T2 — S2 — T3

STATE	RUNNING			STOPPING	RESTARTING	ABORTING	HOLDING
STEP	S1	S2	S3	1	1	1	1
REQUESTS							
Ready for Transfer	1						
Ready to stop transfer		2					
INTERNAL MODULES							
Accumulator Outlet Valve		Open	Close	Close	Open	Close	Close
ALARMS							
TRANSITIONS							
Transfer initiated	T1						
Transfer complete		T2					
Valves closed			T3	Complete	Complete	Complete	Complete
MESSAGES							
PARAMETERS							
REPORTS							

PERMISSIVES AND INTERLOCKS

CONDITION	TYPE

P: PERMISSIVE; I: INTERLOCK

Phase Template

Figure B.1 shows a phase template using structured text that is based on the following state model. The shaded states are active states in which the phase actually performs work. All other states are final states in which no work is performed by the phase. The program follows.

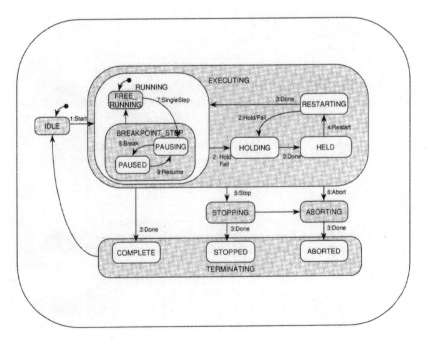

Figure B.1 Phase state model.

```
IF ABORTING THEN (*TEMPLATE FOR 1 STEP ABORT*)
     IF (STEP_NUMBER = 1) THEN (*PERFORM STEP 1*)
          (******PUT THE WORK FOR ABORT STEP 1 HERE *)

          (******PUT THE ABORT COMPLETE CRITERIA HERE *)
          (*IF CONDITIONX THEN*)
          ABORTING_COMPLETE := TRUE; (*SET THE ABORT COMPLETE FLAG*)
          (*ENDIF*)
     END_IF;

ELSIF STOPPING THEN (*TEMPLATE FOR 3 STEP STOP*)
     IF (STEP_NUMBER = 1) AND (NOT PAUSED) THEN (*PERFORM STEP 1*)
          (******PUT THE WORK FOR STOP STEP 1 HERE *)

          (******PUT THE TRANSITIONS OUT OF STEP 1 HERE*)
          (*IF CONDITIONX THEN*)
          STEP_NUMBER := 2;
          (*END_IF*)

          (*THIS PAUSES THE PHASE AFTER THIS STEP IF SINGLE STEP IS ACTIVATED*)
          (******THIS CODE SHOULD BE DELETED IF SINGLE STEP ISN'T ALLOWED AT THIS POINT*)
          IF SINGLE_STEP THEN
               PAUSED := TRUE;
          END_IF;

     ELSIF (STEP_NUMBER = 2) AND (NOT PAUSED) THEN (*PERFORM STEP 2*)
          (******PUT THE WORK FOR STOP STEP 2 HERE *)

          (******PUT THE TRANSITIONS OUT OF STEP 2 HERE*)
          (*IF CONDITIONX THEN*)
          STEP_NUMBER := 3;
          (*END_IF*)

          (*THIS PAUSES THE PHASE AFTER THIS STEP IF SINGLE STEP IS ACTIVATED*)
          (******THIS CODE SHOULD BE DELETED IF SINGLE STEP ISN'T ALLOWED AT THIS POINT*)
          IF SINGLE_STEP THEN
               PAUSED := TRUE;
          END_IF;

     ELSIF (STEP_NUMBER = 3) AND (NOT PAUSED) THEN (*PERFORM STEP 3*)
          (******PUT THE WORK FOR STOP STEP 3 HERE *)

          (******PUT THE STOP COMPLETE CRITERIA HERE *)
          (*IF CONDITIONX THEN*)
          STOPPING_COMPLETE := TRUE; (*SET THE STOP COMPLETE FLAG*)
          (*END_IF*)
     END_IF;

ELSIF HOLDING THEN (*TEMPLATE FOR 3 STEP HOLD*)
     IF (STEP_NUMBER = 1) AND (NOT PAUSED) THEN (*PERFORM STEP 1*)
          (******PUT THE WORK FOR HOLD STEP 1 HERE *)

          (******PUT THE TRANSITIONS OUT OF STEP 1 HERE*)
```

```
(*IF CONDITIONX THEN*)
STEP_NUMBER := 2;
(*END_IF*)

(*THIS PAUSES THE PHASE AFTER THIS STEP IF SINGLE STEP IS ACTIVATED*)
(******THIS CODE SHOULD BE DELETED IF SINGLE STEP ISN'T ALLOWED AT THIS POINT*)
IF SINGLE_STEP THEN
        PAUSED := TRUE;
END_IF;

ELSIF (STEP_NUMBER = 2) AND (NOT PAUSED) THEN (*PERFORM STEP 2*)
        (******PUT THE WORK FOR HOLD STEP 2 HERE *)

        (******PUT THE TRANSITIONS OUT OF STEP 2 HERE*)
        (*IF CONDITIONX THEN*)
        STEP_NUMBER := 3;
        (*END_IF*)

        (*THIS PAUSES THE PHASE AFTER THIS STEP IF SINGLE STEP IS ACTIVATED*)
        (******THIS CODE SHOULD BE DELETED IF SINGLE STEP ISN'T ALLOWED AT THIS POINT*)
        IF SINGLE_STEP THEN
                PAUSED := TRUE;
        END_IF;

ELSIF (STEP_NUMBER = 3) AND (NOT PAUSED) THEN (*PERFORM STEP 3*)
        (******PUT THE WORK FOR HOLD STEP 3 HERE *)

        (******PUT THE HOLD COMPLETE CRITERIA HERE *)
        (*IF CONDITIONX THEN*)
        HOLDING_COMPLETE := TRUE; (*SET THE HOLD COMPLETE FLAG*)
        (*END_IF*)
END_IF;

ELSIF RESTSARTING THEN (*TEMPLATE FOR 5 STEP RESTART*)
    IF (STEP_NUMBER = 1) AND (NOT PAUSED) THEN (*PERFORM STEP 1*)
        (******PUT THE WORK FOR RESTART STEP 1 HERE *)

        (******PUT THE TRANSITIONS OUT OF STEP 1 HERE*)
        (*IF CONDITIONX THEN*)
        STEP_NUMBER := 2;
        (*END_IF*)

        (*THIS PAUSES THE PHASE AFTER THIS STEP IF SINGLE STEP IS ACTIVATED*)
        (******THIS CODE SHOULD BE DELETED IF SINGLE STEP ISN'T ALLOWED AT THIS POINT*)
        IF SINGLE_STEP THEN
                PAUSED := TRUE;
        END_IF;

        (******INSERT THIS CODE IF AFTER THIS STEP IS A PAUSE POINT*)
        (*IF PAUSE_COMMAND THEN
                PAUSED := TRUE;
        END_IF;*)
```

```
ELSIF (STEP_NUMBER = 2) AND (NOT PAUSED) THEN (*PERFORM STEP 2*)
        (*****PUT THE WORK FOR RESTART STEP 2 HERE *)

        (*****PUT THE TRANSITIONS OUT OF STEP 2 HERE*)
        (*IF CONDITIONX THEN*)
        STEP_NUMBER := 3;
        (*END_IF*)

        (*THIS PAUSES THE PHASE AFTER THIS STEP IF SINGLE STEP IS ACTIVATED*)
        (*****THIS CODE SHOULD BE DELETED IF SINGLE STEP ISN'T ALLOWED AT THIS POINT*)
        IF SINGLE_STEP THEN
                PAUSED := TRUE;
        END_IF;

        (*****INSERT THIS CODE IF AFTER THIS STEP IS A PAUSE POINT*)
        (*IF PAUSE_COMMAND THEN
                PAUSED := TRUE;
        END_IF;*)

ELSIF (STEP_NUMBER = 3) AND (NOT PAUSED) THEN (*PERFORM STEP 3*)
        (*****PUT THE WORK FOR RESTART STEP 3 HERE *)

        (*****PUT THE TRANSITIONS OUT OF STEP 3 HERE*)
        (*IF CONDITIONX THEN*)
        STEP_NUMBER := 4;
        (*END_IF*)

        (*THIS PAUSES THE PHASE AFTER THIS STEP IF SINGLE STEP IS ACTIVATED*)
        (*****THIS CODE SHOULD BE DELETED IF SINGLE STEP ISN'T ALLOWED AT THIS POINT*)
        IF SINGLE_STEP THEN
                PAUSED := TRUE;
        END_IF;

        (*****INSERT THIS CODE IF AFTER THIS STEP IS A PAUSE POINT*)
        (*IF PAUSE_COMMAND THEN
                PAUSED := TRUE;
        END_IF;*)

ELSIF (STEP_NUMBER = 4) AND (NOT PAUSED) THEN (*PERFORM STEP 4*)
        (*****PUT THE WORK FOR RESTART STEP 4 HERE *)

        (*****PUT THE TRANSITIONS OUT OF STEP 4 HERE*)
        (*IF CONDITIONX THEN*)
        STEP_NUMBER := 5;
        (*END_IF*)

        (*THIS PAUSES THE PHASE AFTER THIS STEP IF SINGLE STEP IS ACTIVATED*)
        (*****THIS CODE SHOULD BE DELETED IF SINGLE STEP ISN'T ALLOWED AT THIS POINT*)
        IF SINGLE_STEP THEN
                PAUSED := TRUE;
        END_IF;

        (*****INSERT THIS CODE IF AFTER THIS STEP IS A PAUSE POINT*)
```

```
          (*IF PAUSE_COMMAND THEN
                 PAUSED := TRUE;
          END_IF;*)

     ELSIF (STEP_NUMBER = 5) AND (NOT PAUSED) THEN (*PERFORM STEP 5*)
          (******PUT THE WORK FOR RESTART STEP 5 HERE *)

          (******PUT THE RESTART COMPLETE CRITERIA HERE *)
          (*IF CONDITIONX THEN*)
          RESTARTING_COMPLETE := TRUE; (*SET THE RESTART COMPLETE FLAG*)
          (*END_IF*)
     END_IF;

ELSIF RUNNING THEN (*TEMPLATE FOR 10 STEP RUN*)
     IF (STEP_NUMBER = 1) AND (NOT PAUSED) THEN (*PERFORM STEP 1*)
          (******PUT THE WORK FOR RUN STEP 1 HERE *)

          (******PUT THE TRANSITIONS OUT OF STEP 1 HERE*)
          (*IF CONDITIONX THEN*)
          STEP_NUMBER := 2;
          (*END_IF*)

          (*THIS PAUSES THE PHASE AFTER THIS STEP IF SINGLE STEP IS ACTIVATED*)
          (******THIS CODE SHOULD BE DELETED IF SINGLE STEP ISN'T ALLOWED AT THIS POINT*)
          IF SINGLE_STEP THEN
                 PAUSED := TRUE;
          END_IF;

          (******INSERT THIS CODE IF AFTER THIS STEP IS A PAUSE POINT*)
          (*IF PAUSE_COMMAND THEN
                 PAUSED := TRUE;
          END_IF;*)

     ELSIF (STEP_NUMBER = 2) AND (NOT PAUSED) THEN (*PERFORM STEP 2*)
          (******PUT THE WORK FOR RUN STEP 2 HERE *)

          (******PUT THE TRANSITIONS OUT OF STEP 2 HERE*)
          (*IF CONDITIONX THEN*)
          STEP_NUMBER := 3;
          (*END_IF*)

          (*THIS PAUSES THE PHASE AFTER THIS STEP IF SINGLE STEP IS ACTIVATED*)
          (******THIS CODE SHOULD BE DELETED IF SINGLE STEP ISN'T ALLOWED AT THIS POINT*)
          IF SINGLE_STEP THEN
                 PAUSED := TRUE;
          END_IF;

          (******INSERT THIS CODE IF AFTER THIS STEP IS A PAUSE POINT*)
          (*IF PAUSE_COMMAND THEN
                 PAUSED := TRUE;
          END_IF;*)

     ELSIF (STEP_NUMBER = 3) AND (NOT PAUSED) THEN (*PERFORM STEP 3*)
```

```
(******PUT THE WORK FOR RUN STEP 3 HERE *)

(******PUT THE TRANSITIONS OUT OF STEP 3 HERE*)
(*IF CONDITIONX THEN*)
STEP_NUMBER := 4;
(*END_IF*)

(*THIS PAUSES THE PHASE AFTER THIS STEP IF SINGLE STEP IS ACTIVATED*)
(******THIS CODE SHOULD BE DELETED IF SINGLE STEP ISN'T ALLOWED AT THIS POINT*)
IF SINGLE_STEP THEN
      PAUSED := TRUE;
END_IF;

(******INSERT THIS CODE IF AFTER THIS STEP IS A PAUSE POINT*)
(*IF PAUSE_COMMAND THEN
      PAUSED := TRUE;
END_IF;*)

ELSIF (STEP_NUMBER = 4) AND (NOT PAUSED) THEN (*PERFORM STEP 4*)
      (******PUT THE WORK FOR RUN STEP 4 HERE *)

      (******PUT THE TRANSITIONS OUT OF STEP 4 HERE*)
      (*IF CONDITIONX THEN*)
      STEP_NUMBER := 5;
      (*END_IF*)

      (*THIS PAUSES THE PHASE AFTER THIS STEP IF SINGLE STEP IS ACTIVATED*)
      (******THIS CODE SHOULD BE DELETED IF SINGLE STEP ISN'T ALLOWED AT THIS POINT*)
      IF SINGLE_STEP THEN
            PAUSED := TRUE;
      END_IF;

      (******INSERT THIS CODE IF AFTER THIS STEP IS A PAUSE POINT*)
      (*IF PAUSE_COMMAND THEN
            PAUSED := TRUE;
      END_IF;*)

ELSIF (STEP_NUMBER = 5) AND (NOT PAUSED) THEN (*PERFORM STEP 5*)
      (******PUT THE WORK FOR RUN STEP 5 HERE *)

      (******PUT THE TRANSITIONS OUT OF STEP 5 HERE*)
      (*IF CONDITIONX THEN*)
      STEP_NUMBER := 6;
      (*END_IF*)

      (*THIS PAUSES THE PHASE AFTER THIS STEP IF SINGLE STEP IS ACTIVATED*)
      (******THIS CODE SHOULD BE DELETED IF SINGLE STEP ISN'T ALLOWED AT THIS POINT*)
      IF SINGLE_STEP THEN
            PAUSED := TRUE;
      END_IF;

      (******INSERT THIS CODE IF AFTER THIS STEP IS A PAUSE POINT*)
      (*IF PAUSE_COMMAND THEN
```

```
                PAUSED := TRUE;
        END_IF;*)

ELSIF (STEP_NUMBER = 6) AND (NOT PAUSED) THEN (*PERFORM STEP 6*)
        (*****PUT THE WORK FOR RUN STEP 6 HERE *)

        (*****PUT THE TRANSITIONS OUT OF STEP 6 HERE*)
        (*IF CONDITIONX THEN*)
        STEP_NUMBER := 7;
        (*END_IF*)

        (*THIS PAUSES THE PHASE AFTER THIS STEP IF SINGLE STEP IS ACTIVATED*)
        (*****THIS CODE SHOULD BE DELETED IF SINGLE STEP ISN'T ALLOWED AT THIS POINT*)
        IF SINGLE_STEP THEN
                PAUSED := TRUE;
        END_IF;

        (*****INSERT THIS CODE IF AFTER THIS STEP IS A PAUSE POINT*)
        (*IF PAUSE_COMMAND THEN
                PAUSED := TRUE;
        END_IF;*)

ELSIF (STEP_NUMBER = 7) AND (NOT PAUSED) THEN (*PERFORM STEP 7*)
        (*****PUT THE WORK FOR RUN STEP 7 HERE *)

        (*****PUT THE TRANSITIONS OUT OF STEP 7 HERE*)
        (*IF CONDITIONX THEN*)
        STEP_NUMBER := 8;
        (*END_IF*)

        (*THIS PAUSES THE PHASE AFTER THIS STEP IF SINGLE STEP IS ACTIVATED*)
        (*****THIS CODE SHOULD BE DELETED IF SINGLE STEP ISN'T ALLOWED AT THIS POINT*)
        IF SINGLE_STEP THEN
                PAUSED := TRUE;
        END_IF;

        (*****INSERT THIS CODE IF AFTER THIS STEP IS A PAUSE POINT*)
        (*IF PAUSE_COMMAND THEN
                PAUSED := TRUE;
        END_IF;*)

ELSIF (STEP_NUMBER = 8) AND (NOT PAUSED) THEN (*PERFORM STEP 8*)
        (*****PUT THE WORK FOR RUN STEP 8 HERE *)

        (*****PUT THE TRANSITIONS OUT OF STEP 8 HERE*)
        (*IF CONDITIONX THEN*)
        STEP_NUMBER := 9;
        (*END_IF*)

        (*THIS PAUSES THE PHASE AFTER THIS STEP IF SINGLE STEP IS ACTIVATED*)
        (*****THIS CODE SHOULD BE DELETED IF SINGLE STEP ISN'T ALLOWED AT THIS POINT*)
        IF SINGLE_STEP THEN
                PAUSED := TRUE;
```

```
        END_IF;

        (******INSERT THIS CODE IF AFTER THIS STEP IS A PAUSE POINT*)
        (*IF PAUSE_COMMAND THEN
                PAUSED := TRUE;
        END_IF;*)

ELSIF (STEP_NUMBER = 9) AND (NOT PAUSED) THEN (*PERFORM STEP 9*)
        (******PUT THE WORK FOR RUN STEP 9 HERE *)

        (******PUT THE TRANSITIONS OUT OF STEP 9 HERE*)
        (*IF CONDITIONX THEN*)
        STEP_NUMBER := 10;
        (*END_IF*)

        (*THIS PAUSES THE PHASE AFTER THIS STEP IF SINGLE STEP IS ACTIVATED*)
        (******THIS CODE SHOULD BE DELETED IF SINGLE STEP ISN'T ALLOWED AT THIS POINT*)
        IF SINGLE_STEP THEN
                PAUSED := TRUE;
        END_IF;

        (******INSERT THIS CODE IF AFTER THIS STEP IS A PAUSE POINT*)
        (*IF PAUSE_COMMAND THEN
                PAUSED := TRUE;
        END_IF;*)

ELSIF (STEP_NUMBER = 10) AND (NOT PAUSED) THEN (*PERFORM STEP 10*)
        (******PUT THE WORK FOR RUN STEP 10 HERE *)

        (******PUT THE RUN COMPLETE CRITERIA HERE *)
        (*IF CONDITIONX THEN*)
                RUNNING_COMPLETE := TRUE; (*SET THE RUN COMPLETE FLAG*)
        (*END_IF*)
END_IF;
END_IF;
```

Example Product Recipe

Mix into Reactor:

> 100 lbs of Raw Material 1
> 150 lbs of Raw Material 2
> 300 lbs of Raw Material 3

React for 20 minutes by:

> Heating to 150 Deg F
> Continuing to mix
> Adding Catalyst

At same time prepare the distillation column.

When done, transfer the contents through the distillation column at 180 °F.

Send the overheads to Tank 1 and the bottoms to Tank 2. See Figures C.1–C.8.

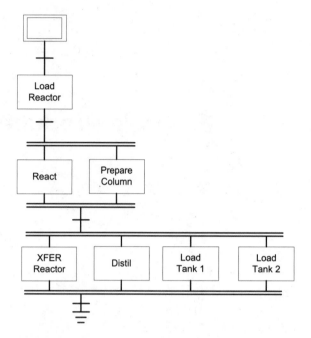

Figure C.1   Example procedure.

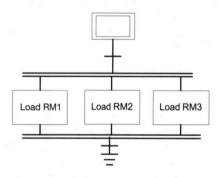

Figure C.2   "Load reactor" operation.

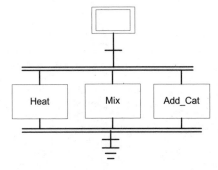

Figure C.3 "React" operation.

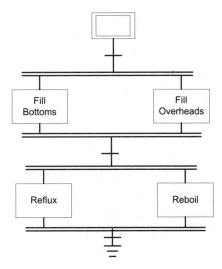

Figure C.4 "Prepare column" operation.

Figure C.5 "XFER Reactor" operation.

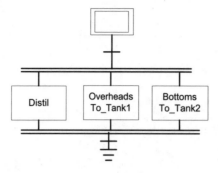

Figure C.6 "Distil" operation.

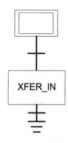

Figure C.7 "Load tank 1" operation.

Figure C.8 "Load tank 2" operation.

D

Qualification Tests

These qualification tests are filled for an example system ABC123.

1. Test Equipment and Calibrations		
Test	**Req'd?**	**Why not?**
1.1. All Sub-Systems	☐yes no	
1.1.1. Test Equipment List w/ Certifications	☐yes no	

2. Site Design and Environmental Conditions		
Test	**Req'd?**	**Why not?**
2.1. Process Automation System ABC123	☐yes no	
2.1.1. Plot Plan Accuracy	☐yes no	
2.1.2. Enclosure Sizing	☐yes no	
2.1.3. Accessibility	☐yes no	
2.1.4. Lighting & Noise	☐yes no	
2.1.5. Physical Protection	☐yes no	
2.1.6. Termination Accessibility	☐yes no	
2.1.7. Communication Cable Lengths	yes no☐	**The Process Automation System ABC123 has no communication cables. No serial communication to/from the Controller is required, per the Design Specifications.**
2.1.8. Temperature & Humidity	☐yes no	

3. Critical Equipment Installation		
Test	**Req'd?**	**Why not?**
3.1. Process Automation System ABC123	☐yes no	
3.1.1. Drawing Verification	☐yes no	
3.1.2. Physical Condition	☐yes no	
3.1.3. Enclosure Seals	☐yes no	
3.1.4. Enclosure Air Flow	☐yes no	
3.1.5. Electrical Classification	☐yes no	
3.1.6. Equipment Records	☐yes no	
3.1.7. Mounting and Labeling	☐yes no	
3.1.8. Spare Parts	☐yes no	
3.1.9. PM Scheduling	☐yes no	
3.1.10. DIP Switches and Jumpers	☐yes no	

4. Power, Grounding and Fusing		
Test	**Req'd?**	**Why not?**
4.1. All Sub-systems	☐yes no	
4.1.1. Power Conditioning	☐yes no	
4.1.2. Power & Ground Wiring	☐yes no	
4.1.3. Ground Condition	☐yes no	
4.1.4. I/O Shield Grounding	☐yes no	
4.1.5. Circuit Breaker Usage	☐yes no	
4.1.6. Circuit Breaker Labeling	☐yes no	
4.1.7. Fuses	☐yes no	
4.1.8. Power Supplies	☐yes no	

5. Wiring, Drawings and I/O Point Lists

Test	Req'd?	Why not?
5.1. All Sub-systems	☐yes no	
5.1.1. Termination Inspection	☐yes no	
5.1.2. Wiring Drawing Accuracy	☐yes no	
5.1.3. Voltage Segregation	☐yes no	
5.1.4. Conduit Junctions	yes no☐	No system conduit spans electrical classification areas. Consequently, conduit junctions do not require special seals.
5.1.5. Communication Link Segregation	yes no☐	The Process Systemprocess automation system has no communication cables. No serial communication to/from the system is required, per the Design Specifications.
5.1.6. Conduits and Cable Trays	☐yes no	
5.1.7. Strain Relief	☐yes no	

6. Field Instrumentation Installation

Test	Req'd?	Why not?
6.1. All Sub-systems	☐yes no	
6.1.1. Device & Instrument Inspection	☐yes no	
6.1.2. Drawing Accuracy	☐yes no	

7. Power-Up and Fault Diagnostics

Test	Req'd?	Why not?
7.1. Process Automation System ABC123	☐yes no	
7.1.1. Startup Procedure	☐yes no	
7.1.2. Status Lights	☐yes no	
7.1.3. Module Failure Lights	☐yes no	
7.1.4. I/O Failure Lights	☐yes no	
7.1.5. Communication Failure Lights	yes no☐	The Process Automation System has no communication failure lights. No serial communication to/from the Controller is required, per the Design Specifications.
7.1.6. Hardware Diagnostics	☐yes no	

8. Loss of Power		
Test	Req'd?	Why not?
8.1. All Sub-systems	☐yes no	
8.1.1. Fail-Safe State (all power off)	☐yes no	
8.1.2. Fail-Safe State (only panel power)	☐yes no	
8.1.3. Fail-Safe State (only field power)	☐yes no	
8.1.4. Startup State (all power on)	☐yes no	

9. Operating System Software Installation		
Test	Req'd?	Why not?
9.1. Process Automation System ABC123	☐yes no	
9.1.1. List Accuracy	☐yes no	
9.1.2. Licenses and Documentation	☐yes no	
9.1.3. Master Storage	☐yes no	
9.1.4. Source Availability	☐yes no	
9.1.5. Installation in System	☐yes no	
9.1.6. Firmware	yes no☐	**Manufacturer's literature indicates that no operating system firmware (EPROM or other) is used in this Controller. Controller operating instructions are an integral part of the hardware and cannot be accessed or modified.**

10. Application Software Installation		
Test	Req'd?	Why not?
10.1. Process Automation System ABC123	☐yes no	
10.1.1. List Accuracy	☐yes no	
10.1.2. File Names	☐yes no	
10.1.3. Master Storage	☐yes no	
10.1.4. Hardcopy	☐yes no	
10.1.5. Installation in System	☐yes no	
10.1.6. Software Diagnostics	☐yes no	

11. Examination of Application Software		
Test	Req'd?	Why not?
11.1. Process Automation System	☐yes no	
11.1.1. Rung and Block Comments	☐yes no	
11.1.2. Dead Code	☐yes no	
11.1.3. Element Identification	☐yes no	
11.1.4. Header	☐yes no	

12. I/O Verification		
Test	Req'd?	Why not?
12.1. Process Automation System ABC123	☐yes no	

1. Operating Parameters		
Test	Req'd?	Why not?
1.1. Process Automation System	☐yes no	

2. Boundary Values		
Test	Req'd?	Why not?
2.1. Process Automation System ABC123	yes no☐	**No operator-adjustable or recipe-adjustable parameters are included in the Process Automation System configuration, per the Design Specifications.**

3. Control Loop/Logic Modules		
Test	Req'd?	Why not?
3.1. Process Automation System ABC123	☐yes no	

4. Operator Interface		
Test	Req'd?	Why not?
4.1. Process Automation System ABC123	yes no☐	No CRT-based operator interface is included in or involved with the Process Automation System, per the Design Specifications.

5. Alarms & Messages		
Test	Req'd?	Why not?
5.1. Process Automation System ABC123	☐yes no	

6. Data Collection & Reporting		
Test	Req'd?	Why not?
6.1. Process Automation System ABC123	yes no☐	No process data is collected or reported by the Process Automation System, per the Design Specifications.

7. Interlocks		
Test	Req'd?	Why not?
7.1. Process Automation System ABC123	☐yes no	

8. Sequence and Ladder Logic		
Test	Req'd?	Why not?
8.1. Process Automation System ABC123	☐yes no	

E

Phase Logic Interface

This is an example of the interface mechanism between a recipe phase and an equipment phase. This information is from SP88.02 Draft 7.

Equipment Control Information Exchange (Recipe to Phase Logic Interface)

This section of this clause defines data structures and communication protocols that can be used to facilitate communications within and between batch control implementations. The communication protocols in this section define the communication requirements between the functions of recipe phase execution and equipment phase execution. This standard may not apply to all batch control applications.

Recipe Phase and Equipment Phase Data Exchange Support

In general, equipment phase logic has control over the events that are performed. The recipe phase will start the equipment phase, but after that the equipment phase logic takes control over most events. The communication interface thus consists of a common state machine enforcement between the recipe phase and equipment phases, a set of services that supports commands to the equipment phase, and a set of services that supports requests from the equipment phase to the recipe phase.

The recipe phase has a number of services, such as arbitration, that it can provide to the equipment phase. These services allow phase logic programmers to write completely stand-alone equipment phases (for example, phases typically don't need to interact within the process control device). This dramatically simplifies the programming effort required for phase logic development. Furthermore, it allows the equipment phases to be resident in multiple process control devices, even those that don't allow for peer-to-peer communication.

The Recipe Phase Expectations of the Equipment Phase

The recipe phase expects equipment phases to conform to specific criteria. The following is an overview of these expectations:

The equipment phase must follow the state transition diagram (STD).

The equipment phase must follow the recipe phase to equipment phase interface protocols.

The equipment phase will use specific requests to access the services provided by the recipe phase.

Function of the Equipment Phase Logic

An equipment phase designed to interact with the recipe phase has some specific requirements that need to be met. The equipment phase logic should be written to implement a user-defined set of sequences for direct control of process equipment to accomplish some process-oriented task (for example, add, mix, or heat). These sequences will perform normal logic as well as abnormal and shut-down logic. Finally, the equipment phase logic should include logic that allows for failure detection and the reporting of the sequence of steps.

Phase State Machine Enforcement

The state machine for the recipe phase to equipment phase interface defines a model for the behavior of these two objects. The purpose of the phase state transition diagram is to ensure that phase logic exhibits the required behavior characteristics to be controllable by the recipe phase.

State Transition Diagram

The state transition diagram (Figure E.1) illustrates the active and quiescent states that are supported by the logic and the paths between these states. The phase logic must adhere to the rules of the state transition diagram. Only valid state transitions as displayed in

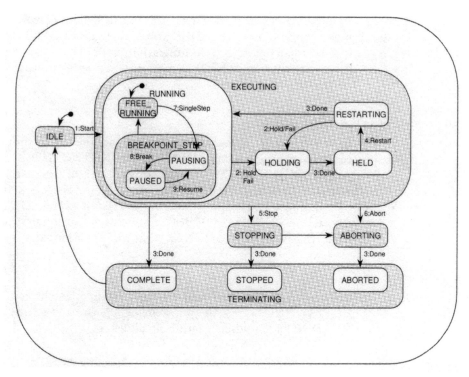

Figure E.1   Phase state transition diagram.

the diagram may be utilized. The configuration applied to the phase logic will vary from site to site, but the logical constructs should conform to the state transition logic. The state transition diagram also shows the states and transitions of the communication protocol between the recipe phase and the equipment phase.

Events Causing State Changes

The following sections describe three kinds of events that can cause transitions between states in the state transition diagram.

Commands

The first type of state transition is referred to as a "commanded" state transition. As shown by the state transition diagram, the phase logic is expected to respond to legal commands according to its current state. For example, when the phase logic is in the IDLE state, a START command should cause the phase logic to transition to the

RUNNING state. If, however, the phase logic receives a STOP command while it is in the IDLE state, this is an invalid command and will be reported as a communication error. The state transition diagram clearly shows what commands are legal in each phase logic state.

Equipment Phase Logic Completion

In the execution of logic, state changes will occur as the result of the completion of one of the "ING" (RUNNING, HOLDING, STOPPING, ABORTING, or RESTARTING) states. Each of these "ING" states is expected to transition to another state when its sequencing has completed. These end states are often referred to as the "ED" (STOPPED, ABORTED, HELD, COMPLETED) states, although the transition from RESTART*ING* to RUNN*ING* is an exception.

Failure Response

A detected failure may cause a state change. In the RUNNING and RESTARTING states, the phase logic is permitted to transition to the HOLDING state when a failure is detected.

Active states

RUNNING

This is the normal active state of the equipment phase. The phase logic is processing its normal execution path. For example, for a material charge phase, the phase logic might be sequencing valves and pumps, and monitoring flowmeters or load cells (see Table E.1).

Note that within the RUNNING state, the phase logic may operate in either a FREE_RUNNING state or a BREAKPOINT_STEP state. When in BREAKPOINT_STEP, the phase logic may be defined to support PAUSED and PAUSING states. PAUSED is a quiescent state. See the BREAKPOINT_STEP command protocol for additional information.

Table E.1 RUNNING state changes

Legal Commands	HOLD	Transition to HOLDING
	STOP	Transition to STOPPING
	ABORT	Transition to ABORTING
Logic Complete		Transition to COMPLETE
Failure Response		Transition to HOLDING

Table E.2 HOLDING state changes

Legal commands	STOP	Transition to STOPPING
	ABORT	Transition to ABORTING
Logic complete		Transition to HELD
Failure response		None

HOLDING

Phase logic is performing necessary logic to transition phase devices to a known state from which batch execution may later be resumed. The actual sequencing necessary may vary depending on the state of the phase's devices and logic when the transition to the HOLDING state occurred (see Table E.2).

RESTARTING

Phase logic is performing necessary logic to transition from a HELD state back to the normal execution path. The actual sequencing necessary may vary depending on how far the phase logic processing has progressed prior to the transition of the phase logic to the HELD state (see Table E.3).

STOPPING

Phase logic is performing necessary logic for a controlled transition from the current state to a known state in which phase logic processing has completed. The actual sequencing necessary to execute the controlled normal transition to the STOPPED state may vary depending on the state of the phase's devices and logic when the transition to the STOPPING state occurred (see Table E.4).

ABORTING

Phase logic is performing necessary logic for an abnormal transition from the current state to a known state in which phase logic processing has completed. The actual sequencing necessary to execute the

Table E.3 RESTARTING state changes

Legal commands	HOLD	Transition to HOLDING
	STOP	Transition to STOPPING
	ABORT	Transition to ABORTING
Logic complete		Transition to RUNNING
Failure response		Transition to HOLDING

Table E.4 STOPPING state changes

Legal commands	ABORT	Transition to ABORTING
Logic complete		Transition to STOPPED
Failure response		None

controlled normal transition to the ABORTED state may vary depending on the state of the phase's devices and logic when the transition to the ABORTING state occurred (see Table E.5). The ABORTING logic is generally expected to perform as rapid as possible a transition to the ABORTED state, differing from the STOPPING logic, which may be expected to perform a more orderly shutdown.

Quiescent States

IDLE

In the idle state, the equipment phase is dormant and inactive. IDLE is considered a "safe" state. IDLE is expected to be the default initial state of a phase (see Table E.6).

COMPLETE

This is the normal terminal state of the phase. COMPLETE is considered to be a "safe" state. Phase logic processing has been completed, and the phase is ready to be reset for its next execution (see Table E.7).

HELD

Phase logic has transitioned to a HELD state in the middle of phase logic execution from which batch execution may later be resumed.

Table E.5 ABORTING state changes

Legal commands	None
Logic complete	Transition to ABORTED
Failure response	None

Table E.6 IDLE state changes

Legal commands	START	Transition to RUNNING (if no failure present)
Logic complete		N/A
Failure response		None

Table E.7 COMPLETE state changes

Legal commands	START	Transition to IDLE
Logic complete		N/A
Failure response		None

Note that the definition of the HELD state does not infer that nothing is happening (see Table E.8). The definition of this state may require that functions such as agitation, temperature control, etc. continue, due to conditions such as material in the tank, etc.

STOPPED

This is an abnormal terminal state of the phase. STOPPED is considered to be a "safe" state. Phase logic processing has been completed, and the phase is ready to be reset for its next execution (see Table E.9).

ABORTED

This is an abnormal terminal state of the phase. ABORTED is considered to be a "safe" state. Phase logic processing has been completed, and the phase is ready to be reset for its next execution (see Table E.10).

Mode Enforcement

The purpose of breakpoint step functionality is to allow the user to force equipment phase execution to stop at programmed "break-

Table E.8 HELD state changes

Legal commands	RESTART	Transition to RESTARTING (if no failure present)
	STOP	Transition to STOPPING
	ABORT	Transition to ABORTING
Logic complete		N/A
Failure response		None

Table E.9 STOPPED state changes

Legal commands	RESET	Transition to IDLE
Logic complete		N/A
Failure response		None

Table E.10 ABORTED state changes

Legal commands	RESET	Transition to IDLE
Logic complete		N/A
Failure response		None

points" in the equipment phase sequencing and wait for operator confirmation before proceeding.

The breakpoint step functionality within the equipment phase API was defined to allow users to take advantage of programmed optional breakpoints in the phase's sequencing logic. The actual programming of breakpoints into the phase's sequencing logic is a completely optional feature that an individual project application team may choose to implement or ignore.

Generally, breakpoint step functionality is useful during the startup/commissioning phase of a project. It is less common, although not unheard of, for a project to use breakpoint step functionality as an optional part of normal batch execution.

When the breakpoint step bit is on, indicating that the phase is in breakpoint step operation, the equipment phase sequencing will pause at every programmed breakpoint in the equipment phase. When a sequence breakpoint is reached, the equipment phase halts sequencing and sets the PAUSED flag to indicate that the equipment phase execution has suspended. The user may then issue a RESUME command to trigger the equipment phase to resume execution until the next breakpoint is reached, or may issue a breakpoint step command, to toggle the equipment phase completely out of breakpoint step mode, triggering the resumption of normal execution.

The sequence breakpoints may or may not be "safe points" in the equipment phase. In most cases, these breakpoints are not process safe points, as they are intended for startup/commissioning checkout of the equipment phase, not normal operation.

Because the process breakpoints are not required to be safe points in the process sequencing, there is the possibility that an operator could put a process into an unsafe condition by placing the equipment phase into breakpoint step execution. Because of this concern, access to the breakpoint step command should have configurable security access.

Breakpoint Step State Machine

Free-Running State

The free-running state is the default initial state of the equipment phase's breakpoint step state machine (see Figure E.2). In this state, the equipment phase sequencing proceeds normally. The sequencing does *not* pause at programmed breakpoints. The values of both the breakpoint step and paused variables contain zero.

Breakpoint Step Transition

The receipt of a breakpoint step command while in the FREE RUN-NING state will cause the equipment phase's breakpoint step state machine to transition to the PAUSING state. The receipt of a break-point step command while in the PAUSING state causes a transition of the state machine to the FREE RUNNING state. The receipt of a breakpoint step command while in the PAUSED state causes a transition of the state machine to the FREE RUNNING state.

Pausing State

The PAUSING state indicates that the equipment phase is executing its sequence logic. However, it will pause execution when it reaches the next programmed breakpoint in the sequencing logic. In this state, the breakpoint step variable has a value of one, whereas the paused variable has a value of zero.

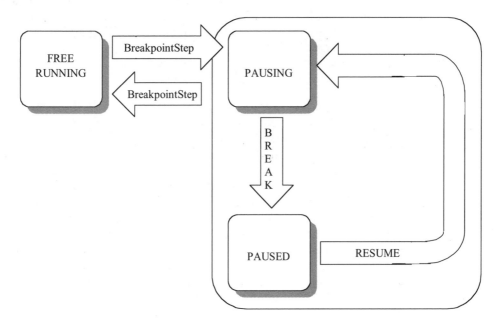

Figure E.2 Breakpoint step state machine.

Break Transition

When the sequencing logic reaches a programmed breakpoint and the current state is PAUSING, the equipment phase transitions to the PAUSED state.

Paused State

The PAUSED state indicates that the equipment phase has stopped at a programmed breakpoint in the equipment phase and is awaiting a resume command from the operator. In the PAUSED state, both the breakpoint step and Paused variables have values of one.

Resume Transition

The receipt of a RESUME command while in the PAUSED state will cause the state machine to transition to the PAUSING.

Program/External Service

The purpose of the program/external attribute of a phase is to act as a permissive for batch server control of equipment phase. When equipment phase is permitted to be controlled by the recipe phase, a zero value should be stored into the owner variable. When equipment phase is not permitted to be controlled by the recipe phase, a nonzero value should be stored into the owner variable. Equipment in EXTERNAL phase should not be able to return to PROGRAM control unless it is in the IDLE state.

Watchdog Service

The purpose of the watchdog protocol is to verify the presence of communication between the recipe phase and the equipment phase. If the communication is deemed to be not present, then both the recipe phase and equipment phase are expected to take preventative action by transitioning any active batch elements to the HELD state.

The watchdog protocol is a simple set/reset protocol. The recipe phase "sets" the watchdog bit by writing a one to the watchdog item. The process connected device (PCD) being accessed is expected to reset the watchdog register to zero whenever a nonzero value is detected.

The recipe phase can detect the presence of a healthy process-connected device by the resetting of the watchdog register. The process connected device can detect the presence of a healthy recipe phase by the periodic setting of the watchdog register.

If the process-connected device determines that the recipe phase has not written to the watchdog register for an unacceptably long period of time, then it is expected to assume that the recipe phase is no longer present and respond by holding all active phases within itself.

If the recipe phase detects that the process-connected device is not resetting the watchdog bit, then it assumes that communication to all variables defined in the data server are not reliable. Batches using any of these variables will be held in response to the communication problems.

Watchdog State Machine

The recipe phase uses the state machine in Figure E.3 to determine the state of every process connected device with which it is communicating:

Lost State

This is the initial state of the recipe phase's conversation status with the process-connected device. In the LOST state, no connection exists between the recipe phase and the process-connected device.

Connect Transition

A successful PCD connection will cause the conversation status to transition from the LOST state to the BAD state. Note that the con-

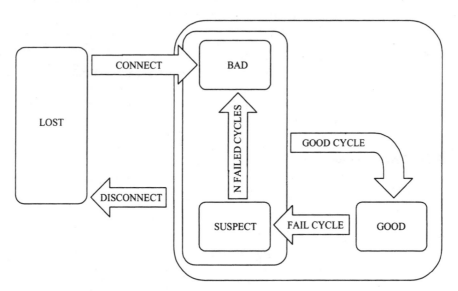

Figure E.3   Watchdog state machine.

versation transitions to BAD instead of GOOD, because we cannot assume that the watchdog logic is functioning correctly just because we got an initial connection.

Bad State

In the BAD state, the recipe phase is not observing successful watchdog cycles with the PCD. Communication to all variables configured to be inside of this PCD are considered unreliable.

Disconnect Transition

A disconnect will cause a conversation in the BAD state to transition to the LOST state.

A disconnect will cause a conversation in the GOOD state to transition to the LOST state. Any batches using variable(s) associated with the conversation will hold, because communication with the variable(s) has been lost.

A disconnect will cause a conversation in the SUSPECT state to transition to the LOST state. Any batches using variable(s) associated with the conversation will hold, because communication with the variable(s) has been lost.

Good Cycle Transition

A single successful watchdog cycle will transition a conversation in the BAD state back to the GOOD state.

Good State

In the GOOD state, the most recent watchdog cycle was completed successfully. Communication with all variables configured to be inside of the PCD associated with this conversation are assumed to be reliable.

Failed Cycle Transition

A single failed watchdog cycle will transition a conversation in the BAD state to the SUSPECT state.

Suspect State

In the SUSPECT state, one or more consecutive watchdog cycles have failed. The number of failed cycles has not yet reached the limit specified in the engineered value WatchdogCycles. Communication with all variables configured to be inside of the PCD associated with this conversation is still assumed to be reliable. The purpose of the suspect state is primarily to indicate when there may be problems with the communication, due to processor loading or other problems.

N Failed Cycles Transition

When the number of consecutive failed watchdog cycles reaches the value defined in the WatchdogCycles value, a conversation in the SUSPECT state will transition to the BAD state. Any batches using variable(s) associated with the conversation will hold, because communication with the variable(s) is no longer considered reliable.

Command Services

The purpose of the command service is to provide transaction assurance in the commands issued to an equipment phase. The command service must guarantee that a new command cannot overwrite a currently pending command value before the equipment phase has a chance to process the pending command.

Many process control devices work on a periodic scan. The recipe phase issues commands to phase logic by writing to the phase's command variable. If the recipe phase were to attempt to issue two or more commands in rapid succession, there would be no mechanism to prevent the recipe phase from overwriting command values before the process control device had an opportunity to "see" them.

To guarantee that a process-connected device "sees" every command issued to the equipment phase by the recipe phase, the recipe phase command handshake protocol was defined. This protocol is used to guarantee that the equipment phase get the opportunity to "see" each recipe phase command as it is issued.

New Parameter Services

The purpose of the new parameters command is to notify a phase that new recipe parameter values are available. This may occur as a result of an operator changing a parameter during processing. When a phase receives a new parameters command, it should respond by making a 1000 series request to request a download of the new parameters.

It is the responsibility of the equipment phase to respond to the new parameters command by requesting a download of the new parameters. This can be done using a normal 1000 series parameter download request.

If a phase will *never* be used in a transfer of control situation, it is possible that support for the new parameters command could be removed from the equipment phase, but remember, *never* is a *very long time.*

Equipment Phase Request Services

There are numerous functions that may be performed by the recipe phase at the request of equipment phase. These functions include arbitration, messaging, reporting, etc. The equipment phase makes a request to the recipe phase by writing a request value into the request variable. Additional parameters describing the request may also be required and are written into the request parameter variables.

The purpose of the request protocol is to control the rate at which the equipment phase makes requests to the recipe phase in order to ensure that the recipe phase detects every request (see Figure E.4).

Dormant State

The request protocol state machine initializes into the DORMANT state. In the DORMANT state, the equipment phase is free to write a request value into the request variable.

Request Transition

The equipment phase makes a request of the recipe phase by writing a request value into the request variable. The act of making a equip-

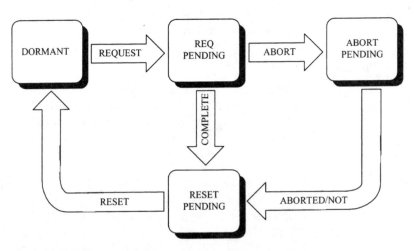

Figure E.4   Equipment phase request protocol state machine.

ment phase request will transition a state machine in the DORMANT state to the REQ PENDING state.

Request Pending State

The REQUEST PENDING (REQ PENDING) state indicates that the equipment phase is waiting for the recipe phase to finish processing an outstanding request. In this state, the equipment phase is not permitted to write to the request register, unless it is attempting to cancel the request currently being processed.

Complete Transition

When the recipe phase has finished processing an equipment phase request, it notifies the equipment phase of the completion of the request by sending a REQUEST COMPLETE command (140) to the equipment phase. When the equipment phase receives this command, it is expected to write an intermediate request value of 10 into the request register. The receipt of this command by the equipment phase will transition a state machine in the REQUEST PENDING state to the RESET PENDING state.

Reset Pending State

In the RESET PENDING state, the equipment phase is waiting for a "RESET FOR NEXT REQUEST" command to be issued by the recipe phase. In this state, the equipment phase is not permitted to write to the request register.

Reset Transition

When the equipment phase receives a "RESET FOR NEXT REQUEST" command (170) from the recipe phase, it is a signal that the recipe phase has seen the intermediate request value. In response to the command the equipment phase is expected to zero the request register. The reception of this command will transition a state machine in the RESET PENDING state to the DORMANT state.

Cancel Transition

If the equipment phase decides that it does not need an outstanding request to be completed, the equipment phase may initiate the canceling of the request by writing a 6000 into the request register. The act of canceling an outstanding equipment phase request will transition a state machine in the REQ PENDING state to the CANCEL PENDING state.

Cancel Pending State

The CANCEL PENDING state indicates that the equipment phase is waiting for the recipe phase to finish processing the cancel of an outstanding equipment phase request. In this state, the equipment phase is not permitted to write to the request register.

Canceled/Not Transition

When the recipe phase finishes processing a request to cancel an outstanding request, it will issue one of two commands to the equipment phase. A "REQUEST CANCELED" command (150) will be issued if the command was successfully canceled. A "REQUEST NOT CANCELED" command (160) will be issued if the recipe phase was unable to cancel the request. In response to a 150 command, the equipment phase is expected to write an intermediate request value of 20 into the request register. If the command is a 160, the equipment phase is expected to write an intermediate request value of 30 into the request register. The reception of either command by the equipment phase will cause the transition of a state machine in the CANCEL PENDING state to the RESET PENDING state.

Failure Notification Service

The purpose of the failure protocol is to provide a mechanism by which the equipment phase can report failures to the recipe phase and through which the recipe phase can respond. Equipment phase is expected to exhibit failure behavior as depicted in the equipment phase failure state machine that is shown in Figure E.5.

Normal State

The NORMAL state indicates that the equipment phase has no detected failures. In this state, the current value of the failure variable for the phase should be zero.

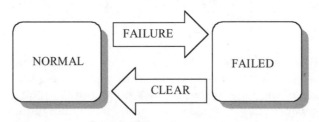

Figure E.5 Equipment phase failure state machine.

Failure Transition

If a failure occurs while the phase is in the NORMAL state, the equipment phase transitions to the FAILED state and indicates the failure by storing a value between 1 and 32767 into its Failure variable.

Failed State

The FAILED state indicates that the equipment phase has detected one or more failures. In this state, the failure variable should have a value between 1 and 32767, indicating the most severe failure that the equipment phase has detected. Higher numeric values indicate more severe failures.

Clear Failures Transition

When a clear failures command is received while the phase is in the failed state, the equipment phase should check to see if any failure conditions still exist. If no failure conditions exist, the equipment phase may transition back to the normal state by storing a value of zero into its failure variable.

Register Exchange Protocol

Phase Communication Variables

A user defines the phases that exist in the plant using a process engineering function. When a phase is defined, the user must define the set of variables to use to communicate with the phase. This set of variables is the "window" through which the recipe phase and the equipment phase communicate. These variables are defined as shown in Table E.11. This set of variables provides the two-way communication interface used for interaction between the recipe phase and the equipment phase.

State Machine Enforcement—
Register Protocol

Most of the recipe phase Commands are to enforce the phase state machine. A recipe phase may command an equipment phase to go to a particular state, as defined by the state transition diagram. Table E.12 describes the commands that may be issued from the recipe phase, including the corresponding values that the recipe phase writes to the equipment phase.

Table E.11 Variables supporting commands

Variable	Description	Write/Subscribe	Use
PHASE_OC:	Command	Write/Subscribe	Send commands to phase logic
PHASE_ST:	State	Subscribe	Indicates the current state of the phase logic
PHASE_RQ:	Request	Subscribe	Make a request to the recipe phase
PHASE_Q##:	Request Params	Write/Subscribe	Additional parameters for defining a request
PHASE_W:	Owner	Subscribe	Permissive for the recipe phase control of phase logic
PHASE_F:	Failure	Subscribe	Indicates failure detected by phase logic
PHASE_PD:	Paused	Subscribe	Indicates that phase logic is paused
PHASE_SS:	Breakpoint step	Subscribe	Indicates that phase logic is in breakpoint step
PHASE_SI:	Step index	Subscribe	Indicates current step index of phase logic
PHASE_UN:	Unit	Write/Subscribe	Indicates the unit with which the phase logic is associated
PHASE_P##:	Parameter	Write/Subscribe	Parameter values for phase logic control
PHASE_R##:	Report	Subscribe	Report value generated by phase logic control

Table E.12 Recipe phase commands

Command	Value	Description
ABORT	10	The ABORT command is sent by the recipe phase to move the phase logic to the ABORTING state. This command is valid in every phase state except IDLE, COMPLETE, STOPPED, and ABORTED. If the phase is in any other state, it will transition to the ABORTING state and execute the ABORTING logic.
HOLD	20	The HOLD command is sent by the recipe phase to move the phase to the HOLDING state. This command is only valid when the phase is RUNNING, RESTARTING, or HELD (conditionally, see _RE).
STOP	30	The STOP command is sent by the recipe phase to move the phase to the STOPPING state. This command is only valid when the phase is in one of the following states: RUNNING, HELD, HOLDING, or RESTARTING.
RESET	40	The RESET command is sent by the recipe phase to move the phase to the IDLE state. This command is only valid when the phase is in one of the following states: STOPPED, COMPLETE, or ABORTED.
BREAKPOINT_STEP	60	The BREAKPOINT_STEP command is sent by the recipe phase to toggle the phase between FREE_RUNNING and BREAKPOINT_STEP. This command causes the phase to set the breakpoint step attribute that allows the phase to transition to the PAUSED state at each programmed pause transition and wait for the operator to issue a RESUME command. The BREAKPOINT_STEP command is valid only when the phase is programmed to respond to this command.
DOWNLOAD	70	The DOWNLOAD command is sent by the recipe phase to indicate to that the recipe phase has crossed a recipe transition (transfer of control). This command triggers the phase to request a download of parameters. Upon execution of this command, the phase sets a Boolean value (_DL) that can be used to initiate a download parameters request. The DOWNLOAD command is valid only when the phase is programmed to respond to this command. Other phases will merely reset the command word without responding in any manner. Typically, this command is used for nonterminating phases that may require setpoint changes during the execution of a batch, such as temperature control, agitation, etc.

(continues)

Table E.12 Recipe phase commands *(continued)*

Command	Value	Description
RESUME	80	The RESUME command is sent by the recipe phase to the equipment phase to resume after the phase has paused at a preprogrammed transition. This command orders a PAUSED phase to resume execution. It is also used to move the phase to the next step during BREAKPOINT_STEP processing. This command is only valid when the PAUSED attribute is active.
RESTART	90	The RESTART command is sent by the recipe phase to the equipment phase to restart the phase logic after it has HELD. The phase logic will transition from the HELD state through the RESTARTING state to the RUNNING state. This command is only valid when the phase is in the HELD state.
START	100	The START command is sent by the recipe phase to the equipment phase to start the phase logic. This command orders the phase to transition from the IDLE state to the RUNNING state. This command is only valid when the phase is in the IDLE state.
CLEAR_FAILURE	130	The CLEAR_FAILURE command is sent by the recipe phase to the equipment phase to clear any outstanding phase failures.
REQ_COMPLETE	140	The REQ_COMPLETE command is sent by the recipe phase to indicate to the equipment phase that the current equipment phase request has been successfully completed. The equipment phase responds by setting the REQUEST register to 10.
REQ_ABORTACK	150	The REQ_ABORTACK command is sent by the recipe phase to indicate to the equipment phase that previous request was successfully aborted. The equipment phase responds by setting the REQUEST register to 20.
REQ_ABORTFAIL The	160	The REQ_ABORTFAIL command is sent by the recipe phase to indicate to the equipment phase that previous request was unsuccessfully aborted. equipment phase responds by setting the REQUEST register to 30.
REQ_RESET	170	The REQ_RESET command is sent by the recipe phase to indicate to the equipment phase that the request has been fully completed and can be reset to allow for a new request. The equipment phase responds by setting the REQUEST register to 0.

Variables Involved

PHASE_ST

The phase logic indicates its current state by storing an integer value into its state variable. The possible state values and their meanings are shown in Table E.13.

Table E.13 States

ABORTING:	10	HOLDING:	20
STOPPING:	30	RESTARTING:	40
RUNNING:	50	HELD:	60
COMPLETE:	70	STOPPED:	80
ABORTED:	90	IDLE:	100

PHASE_OC

The command variable (in conjunction with the command handshake) is used by the recipe phase to issue commands to the equipment phase. The relevant commands for the phase logic state transition diagram are shown in Table E.14.

Breakpoint Step Enforcement—Register Protocol

Variables Involved

PHASE_OC

The command variable (in conjunction with the command handshake) is used by the recipe phase to issue commands to the equipment phase. The relevant commands for the breakpoint step functionality are:

BREAKPOINT_STEP   60
RESUME   80

PHASE_SS

The breakpoint step variable is used by the equipment phase to indicate if the phase is in breakpoint step operation. Breakpoint step operation means that the equipment phase will pause at every programmed breakpoint in the phase's sequencing logic. A value of one indicates that the phase is in breakpoint step operation. A value of zero indicates that it is not.

PHASE_PD

The paused variable is used by the equipment phase to indicate that the phase has paused at a programmed "breakpoint" in the equipment phase and will not continue execution until commanded to resume.

Table E.14 Commands

ABORT:	10	HOLD:	20
STOP:	30	RESET:	40
RESTART:	90	START:	100

Program/External Service—
Register Protocol

The owner attribute of a phase acts as a permissive for control of the equipment phase. When the equipment phase has an owner attribute value of PROGRAM, it is indicating that the recipe phase is permitted to control the equipment phase. When the equipment phase has an owner attribute value of EXTERNAL, the recipe phase is not permitted to attempt to control the equipment phase.

An owner attribute value of PROGRAM is the expected value for normal batch recipe execution. When all phases used by a recipe have an attribute value of PROGRAM, recipe execution can be expected to proceed normally.

If, however, one or more phases used by a recipe have an owner attribute value of EXTERNAL, recipe execution will not proceed normally. There are two scenarios in which a recipe executing in the recipe phase may encounter a phase with an owner attribute of EXTERNAL.

The most common scenario is when the phase has an owner attribute of EXTERNAL prior to the recipe phase attempting to connect to the phase. When a recipe detects that it is attempting to connect to a phase whose owner attribute is EXTERNAL, it generates a failure and holds the batch. The step that attempted to connect to the EXTERNAL phase disconnects and returns to the READY (not connected) state.

The second scenario that may occur is that a phase may transition from PROGRAM to EXTERNAL while it is being used by the recipe phase. When this occurs, a failure is generated and the batch is held. The step that was connected to the phase, however, remains attached to the phase. This attachment is a "view only" attachment. No commands will be sent to the phase, nor will equipment phase requests be processed, but the user will still be able to view the state of the phase on the SFC. Since the phase is under EXTERNAL control, however, the state of the phase is no longer included in the composite state calculations of the higher levels of the batch.

There are several ways in which equipment phase might use the EXTERNAL owner attribute. Some projects provide an alternative operator interface for control of equipment phase. These alternative operator interfaces often exist for use in a failure situation, such as a failure of the recipe phase node. To prevent conflict between the recipe phase and the alternative operator interface, the owner attribute is used.

Some control systems use the owner attribute to indicate that the control hardware is not in the proper state for equipment phase control by the recipe phase.

Variables Involved

PHASE_W

The owner variable is used by the equipment phase to indicate a permissive for the recipe phase control of the equipment phase. The value of this variable is treated as Boolean with zero and nonzero values defined as follows:

The recipe phase control of the equipment phase is permitted (program mode)

$< > 0$ The recipe phase control of the equipment phase is not permitted (external mode)

Watchdog Service—Register Protocol

The engineered WatchdogPeriod controls the frequency at which the recipe phase writes to the watchdog register. In each period, the recipe phase begins by writing a one to the watchdog register. It then waits until the end of the period. If at the end of the period the value of the watchdog register has been reset to zero, then the period is considered to be a successful cycle. If the write fails, or the value of the watchdog register is not zero at the end of the period, then the period is considered to be a failed scan.

Because the recipe phase and the process-connected device independently decide whether to trust the quality of the communication, it is important to see that the time limit tolerances in the recipe phase and the process connected are appropriate in relation to each other.

The recipe phase will not transition a conversation from the GOOD state to the SUSPECT state to the BAD state until the number of consecutive failed watchdog cycles reaches the value specified in WatchdogCycles is met. So, the total length of time before the recipe phase will consider a conversation BAD is the watchdog period multiplied by the tolerance limit for the number of watchdog failures. For example, if the watchdog period was set at 10,000 milliseconds, and the WatchdogCycles parameter was set at 5, then the recipe phase would not change the status of a GOOD conversation to bad until at least 50 seconds (5×10) has passed.

Process connected devices typically program their watchdogs using a single timer that is reset every time a write from the recipe phase is detected. If the timer times out, the communication is considered bad, and all active phases are held.

Generally, we recommend that the tolerance in the process connected device be set a little greater than that in the recipe phase. For example, if the engineered values specified a 10 second period and five timeouts, then the recipe phase would have a tolerance of 50 seconds. In this case, the tolerance in the process-connected device should probably be set to somewhere in the range of 70 seconds.

Variables Involved

WATCHDOG

A watchdog item is configured for each process connected device. The watchdog item is a Boolean value that is set by the recipe phase and reset by the process connected device.

Command Services—Register Protocol

The equipment phase command variable is treated as a 16 bit integer value. When the recipe phase writes a command to a phase command variable, it stores separate values into the upper and lower bytes of the command variable. The low byte is used to hold the command value. Note that this implies that all recipe phase commands must have values of 255 or less. The upper byte is used by the recipe phase to write an "ID" associated with the command. When the equipment phase "sees" the command, it is expected to clear the lower eight bits of the command variable to indicate to the recipe phase that the command has been received and the recipe phase can now issue any new commands.

The clearing of the low byte is simply an acknowledgment that the command has been "seen." It is not intended to be an acknowledgment that the command has been processed! For example, if a STOP command is issued, the handshake (clearing of the low byte) should be done immediately. The user should not wait until the state has changed to STOPPING or STOPPED before completing the handshake. The only purpose of the handshake is to prevent the recipe phase from overwriting a command before the equipment phase has had a chance to see it. If this handshake were not present, it would be possible for serious problems to occur. For example, if two operators almost simultaneously issued ABORT and RESTART commands, and the RESTART command overwrote the ABORT command, the ABORT command would never be seen by the equipment phase.

Command Handshake State Machine and
Transaction Assurance

Figure E.6 depicts the state machine used for the command handshake protocol.

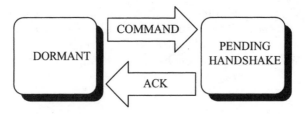

Figure E.6   Command handshake state machine.

Dormant State

The command handshake state machine initializes into the DOR-MANT state. In the DORMANT state, the recipe phase is free to write to the equipment phase's command variable.

Command Transition

The recipe phase issues a command to equipment phase by writing to the command variable. The command value is a 16 bit integer. The low byte of the integer contains the command value and the high byte contains a command ID. The act of writing to the command variable changes the state of the command handshake state machine from DORMANT to PENDING HANDSHAKE.

Pending Handshake State

The recipe phase is not permitted to write to the command variable of a phase whose command handshake state machine is in the PEND-ING HANDSHAKE state. This is to prevent the recipe phase from overwriting a previous command value that the equipment phase has not yet "seen."

Ack Transition

The equipment phase acknowledges that it has "seen" a command by clearing the low byte of the 16 bit command variable. The acknowl-edging of a command changes the state of the command handshake state machine from PENDING HANDSHAKE back to DORMANT.

General Protocol—Commands

The recipe phase uses the high byte of the command variable in a manner that guarantees that a change in value of the command variable will occur. By forcing a change in value, the recipe phase guarantees that it will receive a value update from the data server.

The recipe phase uses the high byte of the command variable to store a "serial number" or command ID on each command. After each command, the value of the command ID is incremented. On startup, the recipe phase checks the current value of the high byte in the command variable to ensure that it will not use a command ID that matches the current high byte value.

Figure E.7 depicts the actual recipe phase command handshake protocol. At time T1, the data server polls the process connected device and reads a value of zero for the command variable. This may or may not be a change from the previous value, but in any event, zero is stored in the data server's database as the latest value of the command variable.

At time T2, the recipe phase writes a 356 value to the command variable for the phase, issuing a START command to the phase. Note that the value 356 is actually 256 + 100, where 256 is the value of one in the high byte, and 100 is the command value in the low byte. The recipe phase will now not issue any commands until it is notified that the value of the low byte of the command variable is zero, indicating that the equipment phase has "seen" the command.

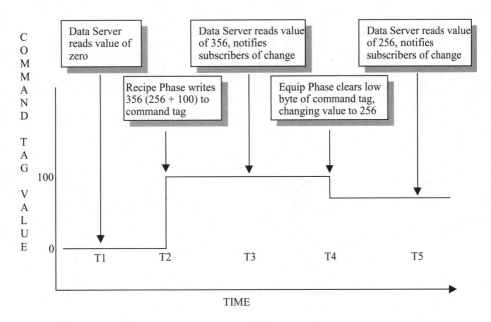

Figure E.7 Command handshake with command ID.

At time T3, the data server polls the process connected device and reads a value of 356 for the command variable. This is a change from the previous value of zero, so the recipe phase, as a subscriber, is notified of the new value of 356 in the command variable. Since the low byte of the command word has not been zeroed, the recipe phase will not issue any new commands to the phase.

At time T4, the equipment phase detects the presence of the command and immediately resets the low byte of the command variable to zero, changing the value of the command variable to 256.

At time T5, the data server polls the process connected device and reads a value of 256 for the command variable. This is a change from the last recorded value of 356, so the recipe phase, as a subscriber is notified of the new value of 256 in the command variable. Since the low byte of the command variable has returned to zero, the command handshake is considered complete, and the recipe phase is free to issue another command.

So, the actual recipe phase command handshake protocol works. Now, let's verify that the protocol still works, even with a data server that polls too slowly to see the value of 356 in the command variable. Consider Figure E.8. At time T1, the data server polls the process-connected device and reads a value of zero for the command variable.

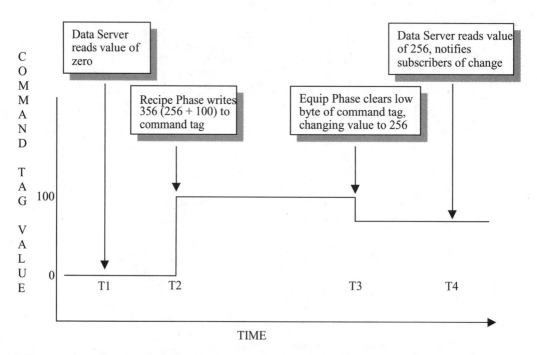

Figure E.8   Command handshake with command ID and slow data server.

This may or may not be a change from the previous value, but in any event, zero is stored in the data server's database as the latest value of the command variable.

At time T2, the recipe phase writes a 356 value to the command variable for the phase, issuing a START command to the phase. Note that the value 356 is actually 256 + 100, where 256 is a value of one in the high byte, and 100 is the command value in the low byte. The recipe phase will now not issue any commands until it is notified that the value of the low byte of the command variable is zero, indicating that the equipment phase has "seen" the command.

At time T3, the equipment phase detects the presence of a command and immediately resets the low byte of the command variable to zero, changing the value of the command variable to 256.

At time T4, the data server polls the process connected device and reads a value of 256 for the command variable. This is a change from the last recorded value of 0, so the recipe phase, as a subscriber, is notified of the new value of 256 in the command variable. Since the low byte of the command variable has returned to zero, the command handshake is considered complete, and the recipe phase is free to issue another command.

By using the high byte of the command word to ensure that the value of the command variable changes, the recipe phase has assured itself that a polling data server, no matter how slow, will see a change in value of the command variable to report to the recipe phase.

Variables Involved

PHASE_OC

The command variable is written to and subscribed on by the recipe phase in support of the command handshake protocol.

New parameter Services—Register Protocol

Variables Involved

PHASE_OC

The command variable (in conjunction with the command handshake) is used by the recipe phase to issue the New Parameters command to the equipment phase. The value of the New Parameters command is:

DOWNLOAD   70

PHASE_RQ

The request variable is used by the equipment phase to make a parameter download request in response to the New Parameters command.

PHASE_Q##

The request parameter variables may be used to provide parameters for the download request made in response to the new parameters command.

Equipment Phase Request Services— Register Protocol

General Protocol—Equipment Phase Request

The actual request protocol is designed to guarantee that the recipe phase detects every equipment phase request, regardless of the poll rate of the data server. The recipe phase request protocol uses a two-step acknowledge. When the recipe phase completes a equipment phase request, a "Request Complete" command is sent to the equipment phase. Rather than zeroing the request parameter in response to the request, however, the equipment phase is expected to write a value of 10 into the request register. The value of 10 is used simply because it is a value that can never be a valid request. The equipment phase then waits for the recipe phase to "see" the value of 10. When the recipe phase sees the value of 10 in the request register, it then sends a "reset for next request" command to the equipment phase. When the equipment phase receives this command, it resets the request register value to zero. The equipment phase can now make another request.

Let us examine this protocol in the time line using Figure E.9. At time T1, the data server polls the process connected device and reads a value of zero for the request variable. This may or may not be a change from the previous value, but in any event, zero is stored in the data server's database as the latest value of the command variable.

At time T2, the equipment phase makes a download parameters request, writing the value 1000 into the request variable.

At time T3, the data server polls the process connected device and reads a value of 1000 for the request variable. This is a change from the previous value of zero, so all subscribers, including the recipe phase, are informed of the new value. The recipe phase, seeing the 1000 value in the request variable, begins processing the parameter download request.

At time T4, the recipe phase has completed the parameter download and sends a "Request Completed" command to the equipment phase. In response to this command, the equipment phase changes the request value register to 10.

At time T5, the data server polls the process connected device and reads a value of 10 for the request variable. This is a change from the

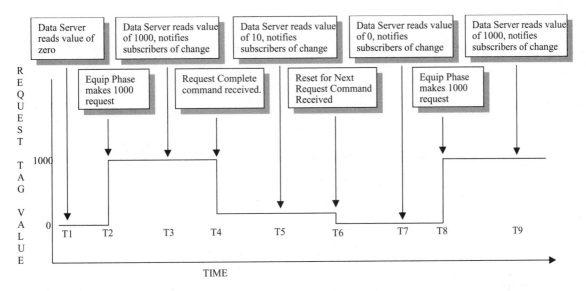

Figure E.9 Equipment phase request protocol with two-step acknowledge.

previous value of 1000, so all subscribers, including the recipe phase, are informed of the new value.

At time T6, in response to seeing a value of 10 in the request register, the recipe phase sends a "Reset for Next Request" command to the equipment phase. In response to this command, the equipment phase changes the request value to zero.

At time T7, the data server polls the process connected device and reads a value of zero for the request variable. This is a change from the previous value of 10, so all subscribers, including the recipe phase are informed of the new value. Since zero is not a valid request, the recipe phase ignores the update.

At time T8, the equipment phase makes a second download parameters request, writing the value 1000 into the request variable again.

At time T9, the data server polls the process connected device and reads a value of 1000 for the request variable. This is a change from the previous value of zero, so all subscribers, including the recipe phase are informed of the new value. The recipe phase, seeing the 1000 value in the request variable, begins processing the new download request.

Now, let us examine the effectiveness of this new protocol when the data server polls too slowly to detect the change in value from 10 to zero to 1000. Figure E.10 depicts this scenario.

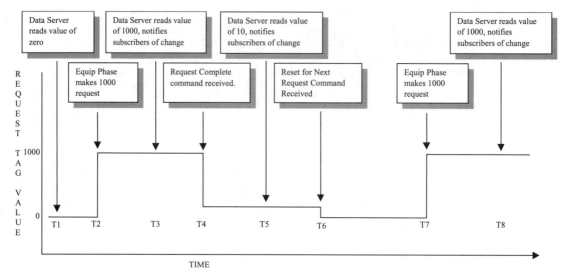

Figure E.10   Two-step acknowledge equipment phase request protocol with slow data server.

At time T1, the data server polls the process connected device and reads a value of zero for the request variable. This may or may not be a change from the previous value, but in any event, zero is stored in the data server's database as the latest value of the command variable.

At time T2, the equipment phase makes a download parameters request, writing the value 1000 into the request variable.

At time T3, the data server polls the process connected device and reads a value of 1000 for the request variable. This is a change from the previous value of zero, so all subscribers, including the recipe phase, are informed of the new value. The recipe phase, seeing the 1000 value in the request variable, begins processing the parameter download request.

At time T4, the recipe phase has completed the parameter download and sends a "Request Completed" command to the equipment phase. In response to this command, the equipment phase changes the request value register to 10.

At time T5, the data server polls the process connected device and reads a value of 10 for the request variable. This is a change from the previous value of 1000, so all subscribers, including the recipe phase, are informed of the new value.

At time T6, in response to seeing a value of ten in the request register, the recipe phase sends a "Reset for Next Request" command to the equipment phase. In response to this command, the equipment phase changes the request value to zero.

At time T7, before the data server has had a chance to poll the process-connected device, the equipment phase makes a download parameters request, writing the value 1000 into the request variable.

At time T8, the data server polls the process-connected device and reads a value of 1000 for the request variable. This is a change from the previously read value of 10, so all subscribers, including the recipe phase, are informed of the new value. The recipe phase, seeing the 1000 value in the request variable, begins processing the new parameter download request.

As you can see from the above analysis, back-to-back identical requests are guaranteed to be detected as unique and separate requests. The recipe phase will detect and process each request, regardless of the poll rate of the data server.

Equipment Request Service—
Variables Involved

PHASE_RQ
The request variable is used by the equipment phase to make requests of the recipe phase and to store intermediate values used by the handshake protocol.

PHASE_Q##
A set of variables used to hold additional parameters for the equipment phase request.

PHASE_OC
The recipe phase uses commands to inform the equipment phase of the completion of requests, the results of an Cancel request, and for handshakes in the request protocol.

PHASE_OC
The command variable is (in conjunction with the command handshake) is used by the recipe phase to issue commands to the equipment phase. The relevant commands for the equipment phase request protocol are:

REQUEST COMPLETED	140
CANCEL SUCCESSFUL	150
CANCEL FAILED	160
RESET FOR NEXT REQUEST	170

MISC

Depending on the actual equipment phase request being made, other variables may be involved. For example, a parameter download request will result in values being written to the parameter variables. For details of the functionality of the individual equipment phase request, please refer to the recipe phase product documentation.

Failure Notification Service— Register Protocol

When the equipment phase detects an error, it stores an error code into its Failure variable. The failure variable value provides several important functions.

First, it provides a mechanism by which an error message can be displayed to the user. Failure values between 1 and 32767 are associated with enumeration's in the PHASE_FAILURES enumeration set. This enumeration set is defined as part of the batch process engineering effort. When equipment phase stores a nonzero value into the failure variable, the recipe phase looks up the string in the PHASE_FAILURES enumeration set. This string can be displayed as an error message. If a value is stored into a phase failure variable for which a string is not defined in the PHASE_FAILURES enumeration set, then the string "CONFIG ERROR - Ordinal not defined in PHASE_FAILURES enumeration set - unable to locate failure description" should be displayed. The failure message for any phase storing a nonzero value into its failure variable should also be recorded as a batch event.

Second, the failure codes allow for display prioritization of multiple failures. The failure codes are prioritized by increasing value. Thus, if two or more phases represented by active steps inside of the same operation have failures, the failure code of the operation will be the highest value of the failure codes of the phases. For example, if one phase had a failure code of 52 and a second phase had a failure code of 8899, the failure code of the operation would be 8899. Thus, if the step representing the operation were selected from the unit procedure SFC, the error message associated with the ordinal 8899 in the PHASE_FAILURES enumeration set could be displayed. This also implies that if a phase detects more than one failure, the failure variable of the phase should be set to the highest value failure that was detected.

Last, but most important, the storing of a nonzero value into a phase's failure variable gives the recipe phase a chance to respond to the failure. Depending upon the way in which the system is defined, some, all, or none of the batch may be held in response to the failure.

Once a failure has been detected, the presence of the failure is effectively "latched" until a Clear Failures command is issued. When the equipment phase receives a Clear Failures command, it should clear the value of the failure register only if no detected failures remain.

Variables Involved

PHASE_OC

The command variable (in conjunction with the command handshake) is used by the recipe phase to issue the Clear Failures command to the equipment phase. The value of the Clear Failures command is:

CLEAR FAILURES   130

PHASE_F

The failure flag is used by the equipment phase to provide indication to the recipe phase of when the equipment phase has detected a failure of some type. A value of zero is recognized as "no failure," and the values 1 through 32767 are recognized as failures with larger numbers representing more severe failures.

Glossary

accident Incident limited to single injury or property damage.

alarm Audible or visible signal that indicates an abnormal or out-of-limits condition.

alarm, absolute An alarm caused by the detection of a variable that has exceeded its high- or low-limit condition.

algorithm A set of well-defined rules that gives a sequence of operations for performing a specific task.

allocation A form of coordination control that assigns a resource to a batch or unit. *Note*—An allocation can be for the entire resource or for portions of a resource.

application Software specifically produced for the functional use of a computer system. Software written to perform a task on the computer.

arbitration A form of coordination control that determines how a resource should be allocated when there are more requests for the resource than can be accommodated at one time.

area A component of a batch manufacturing site that is identified by physical, geographical, or logical segmentation within the site. *Note*—An area may contain process cells, units, equipment modules, and control modules.

as-built The act of updating documentation to reflect how the process or system was actually installed.

basic control Control that is dedicated to establishing and maintaining a specific state of equipment or process condition. *Note*—Basic control may include regulatory control, interlocking, monitoring, exception handling, and xxxxxx xxxxxxx xxxxxxxx.

batch control Control activities and control functions that provide a means to process finite quantities of input materials by subjecting them to an ordered set of processing activities over a finite period of time using one or more pieces of equipment.

batch process A process that leads to the production of finite quantities of material by subjecting quantities of input materials to an ordered set of processing activities over a finite period of time using one or more pieces of equipment.

batch schedule A list of batches to be produced in a specific process cell. *Note*—The batch schedule typically contains such information

353

as what is to be produced, how much is to be produced, when or in what order the batches are to be produced, and what equipment is to be used.

batch   1) The material that is being produced or that has been produced by a single execution of a batch process. 2) An entity that represents the production of a material at any point in the process. *Note*—Batch means both the material made by and during the process and also an entity that represents the production of that material. Batch is used as an abstract contraction of the words "the production of a batch."

baud   A unit of signaling speed equal to the number of discrete conditions or signal events per second.

bottom-up   An approach that starts with the lowest-level components of a hierarchy and proceeds through progressively higher levels to the top level component.

code   To represent data or a computer program in a symbolic form that can be accepted by a processor. Also, one or more computer programs.

common resource   A resource that can provide services to more than one requester. *Note*—Common resources are identified as either exclusive-use resources or shared-use resources.

computer system   A group of hardware components assembled to perform in conjunction with a set of software programs, which are collectively designed to perform a specific function or a group of functions.

consequence   Direct, undesirable result of an accident sequence, usually involving fire, explosion, or release of toxic material.

continuous process   A process in which, for extended periods of time, uninterrupted flows of fluids enter and products leave a system, as opposed to batch process.

control action   Of a controller or of a controlling system, the nature of the change of the output effected by the input.

control, continuous   Control action performed continuously (analog control).

control, cascade   Control in which the output of one controller is the set point for another controller.

control, feedback   Control in which a measured variable is compared to its desired value to produce an actuating error signal that is acted upon in such a way as to reduce the magnitude of the error.

control, feedforward   Control in which information concerning one or more conditions that can disturb the controlled variable is converted, outside of a feedback loop, into corrective action to minimize deviations of the controlled variable.

control, PID   Control action in which the output is proportional to a linear combination of the input, the time integral of input and the time rate-of-change of input

controller A device that operates automatically to regulate a controlled variable.

control loop (loop) A grouping of instruments, control algorithms, and actuators designed to measure and control a controlled variable.

control module The lowest-level grouping of equipment in the physical model that can carry out basic control. *Note*—This term applies to both the physical equipment and the equipment entity.

control recipe A type of recipe which, through its execution, defines the manufacture of a single batch of a specific product.

coordination control A type of control that directs, initiates, and/or modifies the execution of procedural control and the utilization of equipment entities.

device An apparatus for performing a prescribed function.

deviation Any departure from a desired value or pattern.

element A component of a device or a system.

enterprise An organization that coordinates the operation of one or more sites.

equipment control The equipment-specific functionality that provides the actual control capability for an equipment entity, including procedural, basic, and coordination control, that is not part of the recipe.

equipment entity A collection of physical processing and control equipment grouped together to perform a certain control function or set of control functions.

equipment module A functional group of equipment that can carry out a finite number of specific minor processing activities. *Note*—An equipment module is typically centered around a piece of process equipment (a weigh tank, a process heater, a scrubber, etc.). This term applies to both the physical equipment and the equipment entity. Examples of minor process activities are dosing and weighing.

equipment operation An operation that is part of equipment control.

equipment phase A phase that is part of equipment control.

equipment procedure A procedure that is part of equipment control.

equipment unit procedure A unit procedure that is part of equipment control.

exception handling Those functions that deal with plant or process contingencies and other events that occur outside the normal or desired behavior of batch control.

exclusive-use resource   A common resource that only one user can use at any given time.

formula   A category of recipe information that includes process inputs, process parameters, and process outputs.

general recipe   A type of recipe that expresses equipment and site-independent processing requirements.

HAZOP   Hazard and operability study. A rigorous analysis of a process to determine the effect of various conditions.

header   Information about the purpose, source and version of the recipe, such as recipe and product identification, creator, and issue date.

ID   A unique identifier for batches, lots, operators, technicians, and raw materials.

incident   An unplanned event with the potential for undesirable consequences.

interface   A shared boundary. This could be a hardware circuit or software that enables communication.

interlock   a device or control action designed to avoid a hazardous condition by taking a predefined action.

ISA   International Society for Measurement and Control (formerly the Instrument Society of America).

ladder diagram   An industry standard for representing relay control logic.

line; train   See definition for train.

loop, closed   A signal path that includes a forward path, a feedback path, and a summing point and forms a closed circuit.

loop, open   A signal path without feedback.

lot   A unique amount of material having a set of common traits. *Note*—Some examples of common traits are material source, the master recipe used to produce the material, and distinct physical properties.

master recipe   A type of recipe that accounts for equipment capabilities and may include process-cell-specific information.

mode   The manner in which the transition of sequential functions are carried out within a procedural element or the accessibility for manipulating the states of equipment entities manually or by other types of control.

modular   Term referring to a subdivision for standard assembly.

network   A series of devices connected by some type of communication medium. A network may be made up of one or more links.

nuisance alarm   An alarm that requires the operator's attention but does not provide meaningful information.

object   A self-contained module consisting of related data and procedures.

operation A procedural element defining an independent processing activity, consisting of the algorithm necessary for the initiation, organization, and control of phases.

operational specification (Op Spec) A document that defines how a process will operate. It is usually associated with an automated process.

path; stream The order of equipment within a process cell that is used, or is expected to be used, in the production of a specific batch.

personnel and environmental protection The control activity that prevents events from occurring that would cause the process to react in a manner that would jeopardize personnel safety and/or harm the environment.

phase The lowest level of procedural element in the procedural control model.

pneumatic control control signals that are transmitted using air pressure.

procedural control Control that directs equipment-oriented actions to take place in an ordered sequence in order to carry out some process-oriented task.

procedural element A building block for procedural control that is defined by the procedural control model.

procedure The strategy for carrying out a process. *Note*—In general, procedure refers to the strategy for making a batch within a process cell. It may also refer to a process that does not result in the production of product, such as a clean-in-place procedure.

process A sequence of chemical, physical, or biological activities for the conversion, transport, or storage of material or energy.

process action Minor processing activities that are combined to make up a process operation. *Note*—Process actions are the lowest level of processing activity within the process model.

process cell A logical grouping of equipment that includes the equipment required for production of one or more batches. It defines the span of logical control of one set of process equipment within an area. *Note*—This term applies to both the physical equipment and the equipment entity.

process control The control activity that includes the control functions needed to provide sequential, regulatory, and discrete control and to gather and display data.

process input The identification and quantity of a raw material or other resource required to make a product.

process management The control activity that includes the control functions needed to manage batch production within a process cell.

process operation A major processing activity that usually results in a chemical or physical change in the material being processed and that is defined without consideration of the actual target equipment configuration.

process output   An identification and quantity of material or energy expected to result from one execution of a control recipe.

process parameter   Information that is needed to manufacture a material but does not fall into the classification of process input or process output. *Note*—Examples of process parameter information are temperature, pressure, and time.

process stage   A part of a process that usually operates independently from other process stages and that usually results in a planned sequence of chemical or physical changes in the material being processed.

process validation   establishing documented evidence that provides a high degree of assurance that a specific process will consistently produce a product meeting its predetermined specifications and quality attributes.

process variable   Any variable property of a process.

Process Safety Management (PSM)   The application of management systems to the identification, understanding, and control of process hazards to prevent process-related incidents and injuries.

qualification   Documented verification that a system or subsystem is installed and performs as intended per design intentions throughout representative or anticipated operating ranges.

ratio control   Control that maintains a predetermined ratio between two variables.

recipe   The necessary set of information that uniquely defines the production requirements for a specific product. *Note*—There are four types of recipes: general, site, master, and control.

recipe management   The control activity that includes the control functions needed to create, store, and maintain general, site, and master recipes.

recipe operation   An operation that is part of a recipe procedure in a master or control recipe.

recipe phase   A phase that is part of a recipe procedure in a master or control recipe.

recipe procedure   The part of a recipe that defines the strategy for producing a batch.

recipe unit procedure   A unit procedure that is part of a recipe procedure in a master or control recipe.

ROI (return on investment)   A financial measure of the benefit received by making an investment.

S88 (SP88)   An international standard that defines terminology and models for batch control. The standard can also help in the design of batch processes by characterizing and defining them.

self-tuning controller A controller that uses the technique of automatically updating the controller tuning parameters based on changing process conditions.

sensor An element or device that receives information of one type and either transmits it as is or converts it to another type of information.

sequences The manner in which instructions are organized to be implemented in a device.

sequential control The manner of control in which instructions are set up in a sequence and solved in that sequence during execution.

setpoint The desired value of a controlled variable.

shared-use resource A common resource that can be used by more than one user at a time.

site A component of a batch manufacturing enterprise that is identified by physical, geographical, or logical segmentation within the enterprise. *Note*—A site may contain areas, process cells, units, equipment modules, and control modules.

site recipe A type of recipe that is site-specific. *Note*—Site recipes may be derived from general recipes by recognizing local constraints, such as language and available raw materials.

state The condition of an equipment entity or of a procedural element at a given time. *Note*—The number of possible states and their names vary for equipment and for procedural elements.

stream; path See definition for path.

structured program A program constructed of a basic set of control structures, each one having one entry point and one exit point.

supervisory control Control action in which the control loops operate independently, subject to intermittent corrective action; for example, setpoint changes from an external source.

top-down Pertaining to an approach that starts with the highest-level components of a hierarchy and proceeds through progressively lower levels.

train; line A collection of one or more units and associated lower-level equipment groupings that has the ability to be used to make a batch of material.

tune To manipulate the configuration parameters of a loop in order to optimize its effectiveness.

unit A collection of associated control modules and/or equipment modules and other process equipment in which one or more major processing activities can be conducted. *Note*—Units are presumed to operate on only one batch at a time. Units operate relatively independently of one another. This term applies to both the physical equipment and the equipment entity. Examples of major processing activities are react, crystallize, and make a solution.

unit procedure   A strategy for carrying out a contiguous process within a unit. It consists of contiguous operations and the algorithm necessary for the initiation, organization, and control of those operations.

unit recipe   The part of a control recipe that uniquely defines the contiguous production requirements for a unit. *Note*—The unit recipe contains the unit procedure and its related formula, header, equipment requirements, and other information.

unit supervision   The control activity that includes control functions needed to supervise the unit and the unit's resources.

Bibliography

Application of Safety Instrumented Systems for the Process Industry. ISA draft standard, ISA dSP84.01, 1996.

Bluhm, A. R. A practical guide to software validation. *Pharmaceutical Technology, 13*(11), 32–40, 1989.

Bray, M., and Corsberg, D. Practical alarm filtering. *InTech,* February, 1994.

Bronstein, R. J. The concept of a validation plan. *Drug Information Journal, 20,* 37–42, 1986.

Chapman, K. G. A history of validation in the United States. *Pharmaceutical Technology, 15*(10), 82–96; *15*(11), 54–70, 1991.

Chappell, David. *Understanding ActiveX and OLE.* Microsoft Press.

COM specification. *http://microsoft.com/oledev/olecom/title.html.*

Current Good Manufacturing Practices for Finished Pharmaceuticals. 21 CFR, 211, FDA, 1979.

Dorf, Richard. *Modern Control Systems. Reading,* MA, Addison-Wesley, 1983.

Draft guidelines on Good Automated Manufacturing Practice. GAMP forum. UK Pharmaceutical Industry Computer systems Validation Forum.

EC guide Annex 11 (Computerized systems).

Edworthy, J., Loxley, S., and Dennis, I. Improving auditory warning design: relationship between warning sound parameters and perceived urgency. *Human Factors, 33*(2), 205–231.

FDA. *Technical Report—Software Development Activities, Reference Materials and Training Aids for Investigators.* July 1987.

Fisher, Thomas G. *Batch Control Systems: Design, Application, and Implementation.* Research Triangle Park, NC: ISA, 1990.

Fowler, M., and Scott, K. *UMl Distilled: Applying the Standard Object Modeling Language.*Reading, MA, Addison-Wesley, 1997.

Function Block Application Process, Part1. Fieldbus Foundation, 1994.

Functional Safety of Electric/Electronic/Programmable Electronic Systems. IEC draft 1508, 1996.

GAMP 96, Draft Guidelines(Version 2.0) on Validation of Automated Systems in Pharmaceutical Manufacture. The GAMP Forum (UK PICSV), 1996.

Guideline on General Principles of Process Validation. FDA, 1987.

Guidelines for Safe Automation of Chemical Processes. Center for Chemical process safety, AIChE, 1993.

Hoonhout, H. C. M., and Zwaga, H. J. G. Supervisory control behavior and the implementation of alarms in process control. In *Human Factors in Alarm Design.* Taylor & Francis, 1994.

IEC 1131-3. International Electrotechnical Commission.

ISA. *ISA-dS88.02 Batch Control, Part 2: Data Structures and Guidelines for Languages.* February, 1998.

Jacobson, Ivar, Booch, Grady, and Rumbaugh, Jim. *The Objectory Software Development Process.*Reading, MA, Addison Wesley, 1998.

Microsoft Corporation. *Component Object Model Specification, Part I: COM Introduction,* ver 0.9(draft), October 1995.

Moray, N. The role of attention in the detection of errors and the diagnosis of failures in man–machine systems. In *Human Detection and Diagnosis of System Failures.* Plenum Press, New York, 1980.

Object Management Group. *CORBA: Architecture and Specification.* OMG Publications, 1992.

Object Management Group. *Object Management Guide, ver 2.0.* OMG Publications, 1992.

Orfali, R., Harkey, D., and Edwards, J. *The Essential Distributed Objects Survival Guide.* New York, Wiley, 1995.

PCD Programmer's Technical Reference. PID, Inc, 1998.

PMA Computer System Validation Committee. Validation concepts for computer systems used in the manufacture of drug products. *Pharmaceutical Technology, 10*(5), 24–34, 1986.

Prichard, S., and Haxthausen, N. *Object Oriented Batch Integration.* WBF, 1996.

Process Safety Management of Highly Hazardous Chemicals. OSHA 29 CFR 1910.119, 1992.

Quatrani, Terry. *Visual Modeling with Rational Rose and UML.*Reading, MA, Addison-Wesley, 1997.

Rasmussen, J.. Skills, rules and knowledge; signals, signs, symbols, and other distinctions in human performance models. *IEEE Transactions on Systems and Cybernetics, 13*(3), 1983.

Rogerson, Dale. *Inside COM, Microsoft's Component Object Model.* Microsoft Press.

Rouse, W. B. Models of human problem solving. *Automatica, 19,* 613–625.

Rumbagh, J., Blaha, M., Premerlani, W., Eddy, F., and Lorenson,W. *Object Oriented Modeling and Design.* Englewood Cliffs, NJ, Prentice-Hall, 1991.

Shaw, J. A. Smart alarm systems: Where do we go from here? *InTech,* December, 1985.

Stanton, N., (Ed.). *Human Factors in Alarm Design.* Taylor & Francis, 1994.

Taylor, D. A. *Object -oriented Technology: A manager's Guide.* Reading, MA, Addison-Wesley, 1991.

Validation of Computer-Related Systems, Technical Report-18. *PDA Journal of Pharmaceutical Science and Technology, 49*(1), s1–s17.

Wilhelm, R. G. Introduction to Object-oriented Analysis and Design—workshop. 1993

Zuboff, S. *In the Age of the Smart Machine: The Future of Work and Power.* Basic Books, Inc., New York, 1988.

Index

About the Authors

Darrin W. Fleming is Director of Consulting and Engineering Services for Sequencia Corp. (formerly PID, Inc.), a supplier of process automation solutions providing both products and consulting services. Prior to joining Sequencia, he was an independent automation consultant. His consulting practice was focused on training in batch automation. Prior that, he worked in several different manufacturing plants for Dow Corning Corporation. Mr. Fleming has expertise with many systems and experience with ISO, OSHA–PSM, and FDA pharma-cGMPs. He has been responsible for systems validation within a bulk pharmaceutical chemical plant and is a licensed Professional Engineer in Control Systems Engineering (PE-CSE). He is a member of the SP88 Batch Control Standards Committee through the ISA and the World Batch Forum Program Committee.

Velumani A. Pillai is the Global Technology Manager for Automation at Pharmacia & Upjohn, Inc., where he is responsible for automation strategy, technology selection, and standardization. In his previous role he conceptualized, implemented, and supported process automation systems focused on batch processes. Mr. Pillai is an active member of the SP88 committee. He is a senior member of ISA and a member of ISPE. He has a Master of Science (Technology) degree in instrumentation. Prior to Pharmacia & Upjohn, he worked on automation and system integration projects in the pharmaceutical, consumer products, and petrochemical industries.